Steps Towards an Evolutionary Physics

The Sustainable World

Aims and Objectives

Sustainability is a key concept of 21st century planning in that it broadly determines the ability of the current generation to use resources and live a lifestyle without compromising the ability of future generations to do the same. Sustainability affects our environment, economics, security, resources, health, economics, transport and information decisions strategy. It also encompasses decision making, from the highest administrative office, to the basic community level. It is planned that this Book Series will cover many of these aspects across a range of topical fields for the greater appreciation and understanding of all those involved in researching or implementing sustainability projects in their field of work.

Topics

Data Analysis
Data Mining Methodologies
Risk Management
Brownfield Development
Landscaping and Visual Impact Studies
Public Health Issues
Environmental and Urban Monitoring
Waste Management
Energy Use and Conservation
Institutional, Legal and Economic Issues
Education
Visual Impact

Simulation Systems
Forecasting
Infrastructure and Maintenance
Mobility and Accessibility
Strategy and Development Studies
Environment Pollution and Control
Land Use
Transport, Traffic and Integration
City, Urban and Industrial Planning
The Community and Urban Living
Public Safety and Security
Global TrendsTopics

Main Editor

E. Tiezzi

University of Siena
Italy

V. Popov
Wessex Institute of Technology
UK

H. Sozer
Illinois Institute of Technology
USA

W. Timmermans
Green World Research
The Netherlands

G. Walters
University of Exeter
UK

A.D. Rey
McGill University
Canada

A. Teodosio
Pontificia Univ. Catolica de Minas Gerais
Brazil

R. van Duin
Delft University of Technology
The Netherlands

Steps Towards an Evolutionary Physics

E Tiezzi
University of Siena, Italy

WITPRESS   **Southampton, Boston**

E. Tiezzi

University of Siena, Italy

Published by

WIT Press

Ashurst Lodge, Ashurst, Southampton, SO40 7AA, UK
Tel: 44 (0) 238 029 3223; Fax: 44 (0) 238 029 2853
E-Mail: witpress@witpress.com
http://www.witpress.com

For USA, Canada and Mexico

WIT Press

25 Bridge Street, Billerica, MA 01821, USA
Tel: 978 667 5841; Fax: 978 667 7582
E-Mail: infousa@witpress.com
http://www.witpress.com

British Library Cataloguing-in-Publication Data

A Catalogue record for this book is available
from the British Library

ISBN: 1-84564-035-7
ISSN: 1476-9581

Library of Congress Catalog Card Number: 2005928182

CONTENTS

PREFACE

by Sven Jørgensen

If we, in a few words, give information about the new world picture that has been presented by modern physics during the 20th century we could apply the following three statements.

(1) Everything is relative (Einstein).
(2) Everything is uncertain (Niels Bohr, Heisenberg and Schrödinger) (alternative formulation "God does play dice").
(3) Everything is irreversible (Prigogine).

All three statements have contributed to a completely new perception of the universe, its development and its underlying processes.

The third statement has received less attention in society than the two others; but it has actually changed our world picture may be even more than statements number (1) and (2). As pointed out by Ilya Prigogine, although quantum mechanics and general relativity are revolutionary, they are still descendants of classical dynamics and carry radical negation of the irreversibility of time. Time, biological development, the evolution of the universe, and history can only continue in one direction and they do play dice. We understand now that irreversibility is an absolute prerequisite for the development of ecosystems, the history and the entire evolution. The world is so dynamic and so complex that the same conditions will never occur again, but new possibilities can emerge from the present conditions due to the irreversibility principle.

Enzo Tiezzi's book "Steps Towards an Evolutionary Physics" considers all the three statements in his description of the new world picture, but with a particular focus on irreversibility. The book builds to a great extent on Prigogine's work; but Enzo Tiezzi "has moved further away from thermodynamic equilibrium" by considering the evolution and the biological core process of "growth". Moreover, he has taken a holistic view: focused on the entire evolutionary process and not on the single steps. He has attempted to see the forest through the trees.

As did Prigogine, Enzo Tiezzi is using entropy as the core thermodynamic variable. Entropy *can* be used to describe systems far from thermodynamic equilibrium, provided that it is applied as a non-state function. *Free energy and entropy are not state functions when applied on living organisms or ecosystems.* At death, the organisms lose momentarily a major part of their free energy (eco-exergy) and produce an enormous amount of entropy (Schrödinger would say that they lose negentropy), because the free energy of the information embodied in the genes are no longer applicable. Death is, however, a very important feature for systems very far from thermodynamic equilibrium, because without death, the important elements will not be recycled and evolution would stop due to lack of carbon, nitrogen, phosphorus, sulphur and so on, the elements that are needed for construction of life. Death is a prerequisite for life!

Evolution has been described many times in the literature by giving details about the present species and their ancestors and their ancestor's ancestors and so on, represented as fossils. The evolutionary literature also contains a presentation of the development of the genomes, the selection processes and the steadily changing life conditions, including changes in climate. Enzo Tiezzi's description is in contrast to this reductionistic description of evolution. Superholistic, he describes the mechanism behind evolution – not the steps or single processes – and the natural laws controlling and governing evolution. The mechanism is that an energy flow through a system inevitably will be utilized to bring the system further away from thermodynamic equilibrium – for instance, to bring the system from chaos to order (see the Fifth Step of the book). An inflow of energy to a system is both necessary and sufficient to move the system further away from thermodynamic equilibrium. Without energy input, the system will inevitably go towards thermodynamic equilibrium, which means that the system will have no life, no gradients and be dull and homogenous. If there is energy input, on the other hand, the system has to move further away from thermodynamic equilibrium, when the energy needed for maintenance is covered. Furthermore, every step forward in evolution towards the very complex organization that characterizes the last emerging class of organisms, the mammals, builds on the previously achieved order and organization. Life is only possible if it does not start from zero for every new generation. A mechanism to store the information gained is needed – the genes. The prerequisites for life and the evolution of life are therefore an energy flow (which is fortunately provided by the sun) and a system that makes it possible to store information – the genes. Approximately 4 billion years have been available for the many many steps that encompass the entire evolution. The sun has provided the energy needed for each step to move further away from thermodynamic equilibrium.

The application of physical chemistry or thermodynamics helps us to understand

(1) the processes and development of ecosystems,
(2) the driving force behind evolution,
(3) the self-organizing ability of living nature,
(4) how the energy inflow and the genes work together to make up evolution and
(5) the very conditions determining the origin of life.

Combined, the study of these is called *ecodynamics* by Enzo Tiezzi – a very pertinent designation for this new emerging science. It builds on a holistic world view, thermodynamics and *all* the three statements presented above applied on living nature. Ecodynamics can (and should be able to) explain the diversity of life – the biodiversity, which is rooted in the enormous variability of life conditions in time and space – and the beauties of nature – the colour and pattern of butterflies, the spectacular colour symphony of a temperate forest at fall, the songs of birds at dawn, and many more examples or, as expressed by Enzo Tiezzi in the Sixth Step, the songs and shapes of nature. The enormous variability and beauty of the life forms are a result of Monod's combination of necessity and chance: survival is a "must" if the results (information) already achieved should not be lost, but it is also a question of which organisms are fittest under the steadily changing (randomly) conditions.

What is presented here in the preface is in short the challenges and basic ideas of ecodynamics. I am sure that the reader will find ecodynamics as exciting as I do. So, have a good time with this interesting book about a new emerging science.

Copenhagen, 14th November 2005
Sven Erik Jørgensen

DEDICATION

to Ilya Prigogine[1]

---

[1] "I am sure that Ilya would be very flattered by the fact that you want to dedicate your book to him. Thank you in his name." Maryna Prigogine.

ACKNOWLEDGEMENTS

This book would not have been possible without the helpful work of the Ecodynamics group of Siena University, namely Helen Ampt, Simone Bastianoni, Francesca Ciampalini, Angelo Facchini, Alessandro Galli, Nadia Marchettini, Valentina Niccolucci, Trista Patterson, Federico M. Pulselli, Riccardo M. Pulselli, Roberto Ridolfi, Federico Rossi and Donatella Valacchi.

PROLOGUE

Conservation without evolution is death. Evolution without conservation is madness (Gregory Bateson).

Bateson underlines a fundamental characteristic of biological evolution and the biosphere. Cross fertilization between evolutionary and conservative aspects is a *conditio sine qua non* for life and its maintenance.

The First Principle of Thermodynamics is concerned with conservation, the Second Principle regards evolution. Time, oscillations, instability and chaos are accorded scientific dignity by the Second Principle.

Energy and matter are conservative properties of the biosphere (First Principle of Thermodynamics). Their organization and the information embodied in the history of energy and matter are evolutionary properties: such properties regard the thermodynamics of far-from-equilibrium systems. *At equilibrium, energy and matter are blind; far from equilibrium they begin to see* (Ilya Prigogine).

The whole is more than the sum of its parts (Blaise Pascal).

Pascal's proposition underlines another fundamental characteristic of the biosphere and living systems, tracing the first watershed between material non-living systems and living systems. It is contrary to the proposition of Descartes that scientific knowledge is enhanced the more the whole is divided into parts. Descartes's approach works for material non-living systems but not for living ones.

As pointed out by G.L. Stebbins:

"The systematic ordering of the basic components or units of any structure is correlated with similar orders in other similar structures, enabling the structures to cooperate in specific functions, e.g. the synthesis of sugar by photosynthesis. *Relational order* helps organisms to carry out chemical reactions and to cause coordinated movement of the parts (thermodynamics and kinetics)."

At all levels we observe events associated with emergence of novelty, that we can in turn associate with the creative power of Nature (Ilya Prigogine).

Prigogine's view introduces the transition from space culture to time culture, ascribing the scientific revolution at the end of the 20th century to the emerging properties, events and narrative elements of biological history. Concepts such as *dissipative structures* and *self-organization* have become quite popular.

Cycles, arrow of time and *events* represent a new lexicon for chemistry and physics that finally become evolutionary, taking up the challenge of complexity in an evolving biosphere. Chance plus choice (stochasticity) and the interactions between them are the basis of a new *ecodynamic science*. Quality and quantity are both necessary for the global description of Nature. *Biodiversity* is the result and the salient property of biological evolution.

The adventure of biological evolution is marked by *chance events* and exact *choices*: it is a *stochastic adventure* in the etymological meaning of the word, from the Greek *"stokazomai"* to "aim at the target" in archery. The arrows are distributed in an apparently random way around the bull's eye, but the hand of the archer chooses, as far as they can, the direction of the arrow: the system combines chance with selection. This combination can be internal as well when autocatalytic configurations select among their constituents.

Ecosystems arise and evolve stochastically by co-evolution and self-organization. They are complex systems, the components of which are all interconnected, and they do not obey linear deterministic laws.

As underlined by the palaeontologist Roberto Fondi:

"Cells differ from other physical systems by virtue of the increased complexity inherent in their *epigenetic development*, or in other words due to a series of geneses, each of which creates new structures and new functions. No machine or inanimate physical system can increase its complexity as can the simplest living cell".

This means that the information necessary to assemble proteins, which DNA merely copies, can only result from the increase in complexity inherent in an epigenetic process.

Roberto Fondi adds that it is impossible to interpret the living world as a "mere assemblage of objects dominated by the rigid deterministic dialectic of chance and necessity".

Ecodynamics is a new science aiming to offer a proposal of cross-fertilization between Charles Darwin and Ilya Prigogine.

1
FIRST STEP

ON GIANTS' SHOULDERS: A NEW EPISTEMOLOGICAL INSIGHT

"...nos esse quasi nanos gigantium humeris insidentes, ut possim plura eis et remotiora videre, non utique proprii visus acumine aut eminentia corporis, sed quia in altum subvehimur et extollimur magnitudine gigantea"

"...we are like dwarfs on the shoulders of giants, so that we can see more than they, and things at a greater distance, not by virtue of any sharpness on sight on our part, or any physical distinction, but because we are carried high and raised up by their giant size"
John of Salisbury, *Metalogicon*, III, 4, 1159.

1.1. Ecodynamics

As Kuhn[1] remarked, the passing of time often uncovers anomalies that the existing theory can no longer explain. The divergence between theory and reality can become very great, causing serious problems.

Kuhn likens change of paradigm to a change in "visual gestalt" (marks on paper first seen as a bird are now seen as an antelope) providing a new perspective of the elements of a problem that enables a solution to be found for the first time. In this sense, a change in paradigm is similar to a reorientation of "gestalt".

---

[1] T.S. Kuhn, *The Structure of Scientific Revolutions*, University of Chicago Press, Chicago, 1962.

Jørgensen[2] underlines that:

> "The recent development in systems ecology represents a paradigm shift. The paradigm that is now receding has dominated our culture for several hundred years. It views the universe as a mechanical system composed of elementary building blocks. The new paradigm is based on a holistic world view. The world is seen as an integrated whole and recognizes the fundamental interdependence of all phenomena."

Evolutionary physical chemistry or ecological thermodynamics (*ecodynamics*[3]) is an important step in the direction of this paradigm shift. The main laws of classical physical chemistry have to be revisited at the light of a new time "*gestalt*".

The basic laws of physics from the time of Newton to the present day have been time reversible; on the contrary, reality is constituted by phenomenological aspects characterized by irreversibility of time: macromolecular organization, cellular differentiation, life processes.

The reason for this lies in dynamical interactions that take place in complex systems. The analysis of the reality requires major modification of current physical chemistry equations and theories. It is now clear that complex systems and their behaviour can only be analyzed by means of relations including time as directional factor.

In the framework of evolutionary physics, we would deal with goal functions instead of state functions, with configurations of evolutionary processes instead of static molecular structures. The ecodynamic models have to be based on relations evolving in time; "far from equilibrium thermodynamics" (Prigogine) assumes upon itself the role of foundation of a new description of nature.

As pointed out by Ilya Prigogine:

> "Although quantum mechanics and general relativity are revolutionary, as far as the concept of time is concerned, they are direct descendants of classical dynamics and carry a radical negation of the irreversibility of time.

---

[2] S.E. Jørgensen, Toward a thermodynamics of biological systems, Moen Brainstorming Meeting and *Int. J. Ecodynamics*, **1**, 9–27, 2006.
[3] See the new quarterly *International Journal of Ecodynamics*, WIT Press, Southampton, UK, 2006.

Irreversibility is not related to Newtonian time or its Einsteinian generalization, but to an 'internal time' expressed in terms of relations between the various units of which the system is composed, as are relations between particles."

This simply means that the new science will no longer deal with state functions, but rather with evolving ecodynamic functions and processes.

The role of thermodynamics in scientific thought boils down to defining relations and identifying constraints; thermodynamics is the science of what is possible and is to physics as logic is to philosophy. Entropy is the enigma of thermodynamics because it has the intrinsic properties of time irreversibility, quality and information that other thermodynamic functions lack. This is why entropy is a central concept in biology and ecology: entropy is the basis of *ecodynamics*.

Prigogine gives a clear and simple description of the ecodynamic peculiarities of the biosphere: the biosphere is far from equilibrium because it is characterized by instability, bifurcations and dissipative chaos; time is therefore real and plays a fundamental constructive role.

Far from equilibrium, we witness new states of matter having properties sharply at variance with those of equilibrium states. We must therefore introduce the foundations of irreversibility into our basic description of nature.

Prigogine introduces the concept of the arrow of time to describe irreversible changes. The main issue derived from the *theory of dissipative structures* is that evolution and maintenance of open systems far from equilibrium are possible only if irreversible thermodynamic processes occur. Such processes dissipate energy and matter, increasing entropy in the environment.

The evolutionary process is such that systems become more and more complex and organized. Biological diversity is the product of long-term interactions at a genealogical and ecological level: the genealogical interactions regard the dissipation of entropy by irreversible biological processes; the ecological interactions regard entropy gradients in the environment.

1.2. The giants and the steps

We stand on the shoulders of giants, namely: Werner Heisenberg, Max Planck, Gregory Bateson, Francisco Varela and Ilya Prigogine.

The ambitious goal of this book is to try to build a bridge between the physics and chemistry of non-living systems and the physics and chemistry of living systems or evolutionary physical chemistry, a bridge between Darwin and Prigogine.

The principal point behind this epistemological approach is that the physics and chemistry of living systems (evolutionary physics) have different foundations from the physics and chemistry of non-living systems.

Since no blueprints of the bridge have yet been made, the following chapters merely describe the steps to take.

The main idea is that a change of paradigm (or no paradigm) is necessary for a true evolutionary physics, which would be irreducible with respect to classical physics and chemistry, as it is concerned with systems of a completely different logical type.

Changes in paradigm in these disciplines have been aptly described by Jørgensen and Svirezhev[4]:

> "In addition to physical openness, there is also an epistemological openness inherent in the formal lenses through which humans view reality. Gödel's Theorem[5] published in January 1931, introduces epistemic openness in a very strong way. The theorem requires that mathematical and logical systems (i.e. purely epistemic, as opposed to ontic) cannot be shown to be self-consistent within their own frameworks but only from outside. A logical system cannot itself (from inside) decide on whether it is false or true. This requires an observer from outside the system, and this means that even epistemic systems must be open.
>
> We can distinguish between ordered and random systems. Many ordered systems have emergent properties defined as properties that a system possesses in addition to the sum of properties of the components: the system is more than the sum of its components. Wolfram[6] calls these *irreducible systems* because their properties cannot be revealed by a reduction to some observations of the behaviour of the components. It is necessary to observe the entire system to capture its behaviour because everything in the system is dependent on everything else by direct and indirect linkages. The

[4] S.E. Jørgensen and Y.M. Svirezhev, *Towards a Thermodynamic Theory for Ecological Systems*, Elsevier, Amsterdam, 2004.

[5] K. Gödel, *Collected Works Vol. I*, Oxford University Press, New York, 1986.

[6] S. Wolfram, Cellular automata as models of complexity, *Nature*, **311**, 419–424, 1984 and Computer software in science and mathematics, *Sci. Am.*, **251**, 140–151, 1984.

presence of irreducible systems is consistent with Gödel's Theorem, according to which it will never be possible to give a detailed, comprehensive, complete and comprehensible description of the world. Most natural systems are irreducible, which places profound restrictions on the inherent reductionism of science.

According to Gödel's Theorem, the properties of order and emergence cannot be observed and acknowledged from within the system, but only by an outside observer. It is consistent with the proverb: "You cannot see the wood for the trees", meaning that if you only see the trees as independent details inside the wood you are unable to observe the system, the wood as a cooperative unit of trees. This implies that the natural sciences, aiming toward a description or ordering of the systems of nature, have meaning only for open systems. A scientific description of an isolated system, i.e. the presentation of an algorithm describing the observed, ordering principles valid for the system, is impossible. In addition, sooner or later an isolated ontic system will reach thermodynamic equilibrium, implying that there are no ordering principles, but only randomness. We can infer from this that an isolated epistemic system will always ultimately collapse inward on itself if it is not opened to cross fertilisation from outside. Thomas Kuhn's account of the structure of scientific revolutions would seem to proceed from such an epistemological analogy of the Second Law.

As pointed out by Prigogine[7] "at all levels of nature we see the emergence of "narrative elements" ". *Evolutionary systems do not allow reproducible experiments: they do play dice.* They are irreducible due to their great complexity and are not deterministic systems.

There are many models, which although unable to describe living things are nevertheless useful steps in this direction and deal with the complexity of non-living chemical and physical systems. The first chapters are dedicated to these important steps and findings.

1.3. Far-from-equilibrium thermodynamics

In his autobiography, Nobel prize winner Ilya Prigogine recounts his first studies on far-from-equilibrium phenomena:

---

[7] I. Prigogine, Foreword, in: E. Tiezzi, *The Essence of Time*, WIT Press, Southampton, UK, 2003.

"It is difficult today to give an account of the hostility that such an approach was to meet. For example, I remember that towards the end of 1946, at the Brussels IUPAP meeting, after a presentation of the thermodynamics of irreversible processes, a specialist of great repute said to me, in substance: "I am surprised that you give more attention to irreversible phenomena, which are essentially transitory, than to the final result of their evolution, equilibrium".

He adds:

"As we started from specific problems, such as the thermodynamic signification of non-equilibrium stationary states, or of transport phenomena in dense systems, we were faced, almost against our will, with problems of great generality and complexity, which call for reconsideration of the relation of physico-chemical structures to biological ones, while they express the limits of Hamiltonian description in physics. Indeed, all these problems have a common element: time."

While professor of physical chemistry at the University of Brussels, Ilya Prigogine was awarded the Nobel prize in chemistry in 1977. His studies opened a window on nature. For the first time in the history of physical chemistry, tools, methods, equations and models were developed to describe the essence of the evolutionary properties of nature.

This change in paradigm: (a) implies that the intrinsic irreversibility of time has erupted in the basic equations of chemistry and physics; (b) it sustains Pascal's view that the whole is greater than the sum of its parts, negating the statement of Descartes that the world should be divided into the smallest possible parts in order to understand it; (c) it suggests that form and aesthetics (and hence quality as well as quantity) play a role in the evolution of nature. Nature is therefore conceived as φυσισ (*physis,* the word from which *physics* is derived*)* in the original Greek sense, a nature in which time, relations and aesthetics play a fundamental role.

The epistemological and historical foundations of this last concept (c) are explained by Bateson[8,9] (I) and in relation to Archimedes in Plutarch's *Parallel Lives*[10] (II):

---

[8] G. Bateson, *Mind and Nature: a Necessary Unity*, Dutton, New York, 1979.

[9] G. Bateson, *Steps to an Ecology of Mind*, Chandler Publishing Company, New York, 1972.

(I)   "Modern science takes the anti-aesthetic assumption attributed to Bacon, Locke and Newton to an extreme. This assumption was that all phenomena can and must be studied and evaluated only in quantitative terms. However, the role of form, colour, flavour, sound, scent and beauty was fundamental for biological evolution, and is still fundamental today for a scientific view of complexity. Nature is threatened by the linear, mechanistic, arrogant, crude approach of science at the service of a society that "knows the price of everything and the value of nothing"."

(II)   "Archimedes is known for certain useful discoveries: his famous principle based on the concept of density led to imprisonment of merchants doping gold with other metals and he used solar energy to burn the sails of Roman ships. When asked to write these useful things or to invent others, he replied that he only concerned himself with "fine and beautiful" things."

Pascal's concept (b) brings thermodynamics and evolutionary biology to the centre of research on nature, relegating particle physics, molecular chemistry and molecular biology to the background. Obviously they still have a central role in the study of non-living matter and mechanics. This type of approach was guessed many years ago by Liquori[11], professor in physical chemistry at Rome "La Sapienza" University and often nominated for the Nobel prize in chemistry.

Liquori wrote:

"In the case of proteins, molecules owe their stability to the same intermolecular forces that stabilise crystals: Van der Waals forces and weak electromagnetic interactions. In living organisms, very weak forces hold DNA molecules together, as well as proteins and membranes. *Because they are weak, they allow these structures to change conformation in order to change function.*"

---

[10] Plutarch, *Vite parallele Vol. IV*, Italian translation, UTET, Turin, 1996.
[11] A.M. Liquori, *Etica ed estetica della scienza*, Di Renzo Editore, Roma, 2003.

He added:

> "The path chosen by many physicists, not Schrödinger of course,
> was to try to use quantum mechanics, which is a mistake (except for
> the phenomenon of sight) because among other things, the maths of
> quantum mechanics is unwieldy."

With regard to the problem of time (a), we cite Prigogine[12] and his time
paradox (the Second Step of this book is dedicated to this problem):

> "How is it possible that on one hand the basic equations of
> dynamics, classical or quantum, are time reversible and that at
> microscopic level, on the other hand, the arrow of time plays a
> fundamental role: How can "time" come from "non time"?"

The basic laws of physics from the time of Newton to the present day have
been time reversible; on the contrary, reality has phenomenological aspects
(macromolecular organization, cellular differentiation, life processes)
characterized by irreversibility of time. The reason for this lies in dynamical
interactions that take place in complex systems.

As stated by Prigogine[12]: "Time is not in the molecules, it is not in their
motion. Time is in the relations between the particles".

The key point of Prigogine's idea is that the analysis of reality, for
example the study of the interaction between complex biomolecular systems,
requires major modification of the current equations and theories of physical
chemistry. It is now clear that complex systems and their behaviour can only
be analyzed by means of relations that include time as directional factor.

On one hand, the use of time-reversible classical and quantum physical
chemistry approaches for studying matter at molecular level and the
behaviour of simple molecular systems has greatly improved our
understanding. On the other hand, we need new approaches and new time-
irreversible theories to describe the behaviour of complex systems.

In the foreword of "The essence of time"[13] Prigogine writes:

> "The first part of this book deals with the passage "from a space
> to a time culture". This is indeed an essential part of the scientific

---

[12] I. Prigogine, in: C. Rossi and E. Tiezzi, eds., *Ecological Physical Chemistry*, Elsevier, Amsterdam, 1–24, 1991.

[13] E. Tiezzi, *The Essence of Time*, WIT Press, Southampton, UK, 2003.

revolution we are witnessing at the end of the 20th century. Science is a dialogue with nature. In the past this dialogue has taken many forms. We feel that we are at the end of the period which started with Galileo, Copernicus and Newton and culminated with the discovery of quantum mechanics and relativity. This was a glorious period but in spite of all its marvelous achievements it led to an oversimplified picture of nature, a picture which neglected essential aspects. Classical science emphasized stability, order and equilibrium. Today we discover instabilities and fluctuations everywhere. Our view of nature is changing dramatically. At all levels we observe events associated with the creative power of nature. I like to say that at equilibrium matter is blind, far from equilibrium it begins to "see"."

Faced with the evolutionary character of nature and life, classical science (physics and chemistry) encounters three paradoxes:

- Prigogine's time paradox;
- the paradox of negentropy that cannot be calculated on the basis of conservative, deterministic and purely quantitative terms (energy and classical entropy) but which must consider information, forms and quality;
- the probability paradox (probability is an aseptic, atemporal mechanistic concept) that has to account for events, emerging phenomena, choices made by plants, animals and ecosystems, random fluctuations of evolutionary biology and the phenomena of far-from-equilibrium systems.

These three paradoxes will be discussed in the following Steps.

1.4. A metaphysical design

Before examining the paradoxes, the main concepts valid both for complex non-living and living systems and some new proposals, another epistemological aspect must be considered.

Geological, meteorological, ecological, oceanographic and biological studies demonstrate that the life of every organism is part of a large scale process that involves the metabolism of the whole planet. *Biological activity is a planetary property, a continuous interaction of atmospheres, oceans, plants, animals, microorganisms, molecules, electrons, energy and matter, all part of the whole.* The role of each of these components is essential for the maintenance of life. Morowitz[14] writes that the environment and living organisms are all inseparable parts of the unity of planetary processes. He goes on to say that the prolonged activity of the global biogeochemical system is more characteristic of life than the individual species that appear, prosper for a period and disappear in the course of evolution.

Unity means complexity and complexity is necessary for the life of living systems: simplification means instability, weakened defences, deterioration. The correlations between the constituents of the natural system, its diversification, individuality, and thus complexity, allow it to be more flexible, to adapt to environmental changes, and to have a greater probability of surviving and thus evolving.

Biodiversity and the marvellous biological beauty are in favour of a metaphysical design in the evolution of life.

Recognition of a metaphysical design in nature is not in line with the ideology of creationism, but rather with the Darwinian evolutionary view (but not with his deterministic drift) or that of foundationless evolution in which free will, choice and chance play a wondrous complex game, a metaphor for which could be the tower of Babel.

This point of view is also distant from positivist, neo-Darwinian ideology that explains nature solely in terms of chance and necessity. Materialist and creationist dogmas are merely two sides of the same scientific stupidity.

It was science, in particular Prigogine's thermodynamics, that discovered the narrative beauty of biological evolution, accentuating the evolutionary character of nature. Evolutionary physics, concerned with the study of nature, uses qualities (not only quantities) and time (as did Darwin). In line with Gödel's theorem, it cannot be constrained within a single mathematical structure and it does not envisage experimental reproducibility. As did biological evolution, evolutionary physics seeks recognition and full scientific dignity.

Two aspects in particular support this point of view.

(a)   We are able to imagine large numbers, but nature is in only three dimensions, like our skulls. Our cognitive processes can only be in three dimensions, but we "feel" that there could be others (e.g. time).

(b)   Different life times exist in nature: those of humans, sequoias, moths, and dogs. Lorenz (Nobel laureate in biology, 1973),

---

[14] H. Morowitz, Due punti di vista sulla "vita", *Scienza 83*, **5**, giugno 1983.

taking the philosophy of Nietzsche as starting point, wrote[15]: "When God created the world, He had inscrutable reasons for giving dogs a life one fifth that of their masters", and indeed one day, the first "golden jackal" (as Lorenz scientifically and poetically expressed it) began to follow a human. We know well that historical time is very different from biological time[16].

To aim for an evolutionary view of the Earth also means working towards unification of the two cultures (science and humanities).

The illuminist dream of reason dominating nature, rather than living in harmony with it, has generated the monsters of *one-way thought,* a type of thought that does not observe the times and modes of nature, that does not know its constraints and limits. Yet it is from limits and constraints that creativity springs, artistic and scientific creativity. Freedom is not of this world, it is not part of our nature. Nature is made of spatial and temporal limits and constraints: our life is not eternal, our dimensions are three, our body weight is what it is, likewise our possibilities of movement. We could say that the beauty and diversity of evolutionary history lie in the fact that every living species has different limits and constraints. Some do not walk erect, others only move in water, others fly in the air: biodiversity consists in the fact that every plant, animal and human has different constraints and has to learn to live with them. They are life itself and determine diversity, without which art and science would not exist.

What we have to do is to fuse microscopic and macroscopic, supersede the dichotomies of reductionism and antireductionism, study biological phenomena in terms of relations and self-organization, so that the behaviour of the parts becomes coherent.

Evolutionary physics will break away from the safety of determinism and/or subjectivism and add irreversibility and uncertainty to its basic paradigms. In other words, it will accept the stochastic nature of time as an intrinsic property of matter in a universe in which, to invert Einstein's phrase, God plays dice to give chance to choice.

To quote Ho[17], "science is a quest for the most intimate understanding of nature. It is not industry set up for the purpose of validating existing theories

[15] K. Lorenz, *So kam der Mensch auf den Hund,* Deutscher Taschenbuch Verlag GMBH & KO.KG, München, 1983; Italian translation *E l'uomo incontrò il cane,* Adelphi Edizioni, Milano, 1998.

[16] E. Tiezzi, *The End of Time,* WIT Press, Southampton, UK, 2003.

[17] M.W. Ho, *The Rainbow and the Worm,* World Scientific Publ. Co., Singapore, 1998.

and indoctrinating students in the correct ideologies. It is an adventure of the free, enquiring spirit which thrives not so much on answers as unanswered questions. It is the enigmas, the mysteries and paradoxes that take hold of the imagination, leading it on the most exquisite dance".

If science is this marvellous dance, the challenge is, to day, to aim to a "thermodynamics of organized complexity", having in mind that "there is as yet no science of organized heterogeneity or complexity such as would apply to living systems".

2
SECOND STEP

THE PURE DURATION OF TIME

"The fundamental things apply as time goes by"

Casablanca, 1942

2.1. The essence of time

This chapter assumes that space and time belong to different logical types, the first being reversible and conservative, the second irreversible and evolutionary. Time (t) and space (s) are reciprocally irreducible quantities.

The chapter summarizes the main ideas of "The Essence of Time" (Tiezzi)[1] and refers to the Thermodynamics of the Nobel laureate in Chemistry, Prigogine and Stengers[2], who wrote "Time is no more opposing man to nature but on the contrary marks his belonging to an inventive and creative universe".

The biological complexity that we observe today in the natural world is the result of the temporal and spatial constraints of our planet and a long evolutionary history made up of relations accumulating in time. Different hierarchies of biological complexity have marked different phases of the Earth's history: membranes with selective permeability and active transport of metabolites three billion years ago, differentiated systems of organs and tissues one billion years ago, the central nervous system 600 million years

---

[1] E. Tiezzi, *The Essence of Time*, WIT Press, Southampton, UK, 2003.
[2] I. Prigogine and I. Stengers, *Entre le Temps et l'Eternité*, Fayard, Paris, 1988 and I. Prigogine, *The End of Certainty – Time's Flow and the Laws of Nature*, The Free Press, New York, 1997.

ago, warm blooded animals 150 million years ago, the first hominids three million years ago, the use of tools about 100,000 years ago.

The rules of the system of relations, the underlying hierarchy, the order in space and time and the correlations between time and form all still have to be studied and understood.

In the beginning, a great void, infinite spaces.

Two atoms meet, recognize each other, exchange energy in collision and proceed on different paths. Their meeting caused changes in energy and direction. The new directions embody the information of the event: a collision occurred and the information it left cannot be cancelled. The new direction carries the memory of the collision.

If we consider hydrogen and oxygen, the meeting gives rise to the water molecule, completely changing the history of biological evolution.

Later in the "time of events" there are other collisions and relations: between three, four and more molecules, relations of a higher order, according to the natural order ticked out by the "clock of events". Prigogine[3] speaks of "natural time ordering". Relations between molecules occur as a consequence of previous events: the evolution of events influences subsequent events in a stochastic way.

During its history, the molecule takes different forms; the history of events is transformed into structure, form, aesthetic content, on the basis of chance and choices made by the initial molecules. As we have seen, chance and choice mean stochastic processes; biological evolution is stochastic, the flow of events and forms is stochastic, our learning process is stochastic because even our minds are part of that marvellous history of coevolution that is the history of life and the biosphere.

2.2. Prigogine's time paradox

The equations of classical physics have no notion of the "time of events" or the "time of things". Prigogine and Stengers[2] wrote, as we have seen, that although quantum mechanics and general relativity are revolutionary, as far as the concept of time is concerned, they are direct descendants of classical dynamics and carry a radical negation of the irreversibility of time.

The recognition that time is "real" leads to what Prigogine called the "time paradox". He asked how could the basic equations of classical and quantum

---

[3] I. Prigogine, in: C. Rossi and E. Tiezzi, eds., *Ecological Physical Chemistry*, Elsevier, Amsterdam, 1–24, 1991.

mechanics could be reversible with respect to time at microscopic level, whereas the arrow of time plays a fundamental role at macroscopic level. *How can time emerge from non-time?*

To solve this paradox, Prigogine started with Poincaré's theorem of 1889, introducing the distinction between integrable and non-integrable systems. The latter lead to an alternative formulation of dynamics in probabilistic terms, in classical and quantum physics alike. This description includes the breaking of time symmetry and incorporates the Second Law of Thermodynamics. The new theory on which the Brussels school is working may have important practical applications in ecology. Classical and quantum physics have practically nothing to contribute to the study of complex ecological systems in a process of evolution.

Prigogine considered Large Poincaré Systems (LPSs) including multiple-body systems involving collisions. He treated them in a way that implied the existence of chaos in the context of dynamics, hitherto regarded as the stronghold of deterministic description. Broken time symmetry and irreversibility erupted in the core of dynamics. Prigogine writes of the inversion of the usual formulation of the time paradox. The usual attempt involved the deduction of the arrow of time from dynamics based on reversible time equations. He speaks of generalizing dynamics to include irreversibility. The divergences of Poincaré are eliminated by an appropriate time ordering of dynamic states.

In this way Prigogine introduced the concept of natural time ordering of dynamic states. To understand what this means, let us take the example of a stone that falls into a pond and causes ripples. We may also have the inverse situation in which incoming waves eject a stone. It is true that only the first event ever occurs. Natural time ordering has the falling stone first and then the ripples. In order to give meaning to the Poincaré denominators, the time of dynamic states must also be ordered: first the unstable atomic state, then the emission of radiation. Prigogine went on to say that natural time ordering must also be introduced into statistical descriptions. This brings LPSs into quite general situations and creates a new dynamics that breaks radically with the past.

In a classical gas, particles collide and these collisions give rise to relations. At first we have binary relations, then ternary relations until more and more particles are involved. Prigogine considers the example of two persons in conversation: this can be regarded as a collision. When the persons leave each other, the memory of their conversation remains. The unidirectionality of relationships breaks the symmetry of the classical description: the time of relationships evolves. Transitions involving higher-order relations are "future-orientated"; those involving lower-order relations are "past-orientated". In the old situation, there was microscopic reversibility and macroscopic irreversibility of time. Now we have a new microscopic level with broken time symmetry from which a dissipative macroscopic level emerges.

We cannot stop the flow of relations or the decay of unstable atomic states. These concepts bring us to the threshold of a new physics that will incorporate dynamics, instability, chaos and irreversibility: an evolutionary physics based on the assumptions of Prigogine.

The science of Descartes, Newton and Einstein, having failed to harness time in rational models, solved the problem by eliminating it. Time is only an illusion of the mind, according to Einstein; contrary to all evidence, time, like space, is measured reversibly in Newtonian and quantum mechanics; time that passes has no place in classical science, where two dogmas negate the flow of time.

One is the dogma of reversibility and the other the dogma of the reproducible experiment. In all western schools and universities it is taught that an experiment only has scientific value if it is reproducible, only if the same result can be obtained again later. This may be true for a machine or an inanimate object that only change in the time scale of the geological eras, but it is certainly not true for biological or ecological experiments. Living creatures and ecosystems obey the laws of biological evolution: at any time they are different from what they were an instant before. This is the essence of life. In biology and ecology, reproducible experiments do not exist and for classical science time does not exist.

We can conclude that biological evolution and ecology are not sciences or, more wisely, with Prigogine, that the fact that quantum mechanics cannot tell us the probability that a quantum transition will occur in a given moment, means that this formalism is incomplete. If we intend taking the notion of time, life and the associated Uncertainty Principle seriously, we have to modify the notion of the observable in quantum mechanics and give an intrinsic meaning, independent of the act of observation, to the probabilistic dispersal of energy. In other words, we have to accentuate the probabilistic character of the theory and not the deterministic or subjective character. Time is in matter; it is in the nature of the molecules; it is an integral part of biological evolution.

Prigogine and Stengers[2] add that the theory of relativity, cosmology and quantum mechanics have always sought separation from the time dimension, placing the laws of time in the dimension of eternity. However, our lives are not governed by atemporal and deterministic laws, but are immersed in the flow of time, in constant relation with memory of the past and projection towards the future.

Boltzmann denied irreversibility, regarding it as a defeat; the physicists of Einstein's generation made this negation a scientific dogma. As Poincaré[4] pointed out in 1889, nothing more sophisticated than plain logic is needed to reveal the error of trying to explain irreversibility in terms of the reversible.

We may add that there is also a space paradox: space can be infinite, but it only has three dimensions.

2.3. Bergson

Lucretius[5] wrote that "time *per se* does not exist: the sense of what has been done in the past, what is in the present and what will be is embodied in things themselves."

Bergson[6] wrote that if time were not invention, it was nothing. Creativity is the prime mover of life and biological evolution.

Science has to invent concepts to describe nature. Science invented the concept of time. Time is a scientific concept. The arrow of time (an irreversible vector with only one dimension and only one direction) is its mathematical model.

Mae-Wan Ho[7] aptly described the relations between thermodynamics and Bergson:

"The second law of thermodynamics defines a time's arrow in evolution and in that sense, captures an aspect of experience which we know intuitively to be true. By contrast, the laws of microscopic physics are time reversible: one cannot derive a time's arrow from them, nor are they affected by a reversal of time. There is another sense in which time in physical laws does not match up to our experience. Newtonian time, and for that matter, relativistic time and time in quantum theory, are all based on a homogeneous, linear progression – the time dimension is infinitely divisible, so that

---

[4] H. Poincaré, Sur les tentatives d'explication mécanique des principes de la Thermodynamique, *C.R. Acad. Sci.*, **CVIII**, 550–553, 1889.

[5] Lucretius, Tempus item per se non est, sed rebus ab ipsis consequitur sensus, transactum quid sit in aevo, tumquae res instet, quid porro deinde sequatur, *De rerum natura*, **I**, 459–461.

[6] H.L. Bergson, *Time and Free Will. An Essay on the Immediate Data of Consciousness*, (F.L. Pogson, translation), George Allen & Unwin, Ltd., New York, 1916.

[7] M.W. Ho, *The Rainbow and the Worm*, World Scientific Publ. Co., Singapore, 1998.

spatial reality may be chopped up into instantaneous slices of immobility, which are then strung together again with the 'time line'. Real processes, however, are not experienced as a succession of instantaneous time slices like the successive frames of a moving picture. Nor can reality be consistently represented in this manner.

The mismatch between the quality of authentic experience and the description of reality given in western science has long been the major source of tension between the scientists and the "romantics". But no one has written more vividly on the issue of time than Bergson, with so few who could really understand him. He invites us to step into the rich flowing stream of our consciousness to recover the authentic experience of reality for which we have substituted a flat literal simulacrum given in language, in particular, the language of science.

In the science of psychology, words which express our feelings – love and hate, joy and pain – emptied of their experiential content, are taken for the feelings themselves. They are then defined as individual psychic entities (or psychological states) each uniform for every occasion across all individuals, differing only in magnitude, or intensity. Should we connect our mind to our inner feelings, what we experience is not a quantitative increase in intensity of some psychological state but a succession of qualitative changes which "melt into and permeate one another" with no definite localizations or boundaries, each occupying the whole of our being within this span of feeling which Bergson refers to as "pure duration".

Pure duration is our intuitive experience of inner process, which is also inner time with its dynamic heterogeneous multiplicity of succession without separateness. Each moment is implicated in all other moments. Thus, Newtonian time, in which separate moments, mutually external to one another, are juxtaposed in linear progression, arises from our attempt to externalize pure duration – an indivisible heterogeneous quality – to an infinitely divisible homogeneous quantity. In effect, we have reduced time to Newtonian space, an equally homogeneous medium in which isolated objects, mutually opaque, confront one another in frozen immobility."

Mae-Wan Ho underlines that:

"Bergson's protests were directed against one of the most fundamental assumptions underlying modern western science. It claims to express the most concrete, common-sensible aspect of nature: that material objects have simple locations in space and time. Yet space and time are not symmetrical. A material object is supposed to have extension in space in such a way that dividing the space it occupies will divide the material accordingly. On the other hand, if the object lasts within a period of time, then it is assumed to exist equally in any portion of that period. In other words, dividing the time does nothing to the material because it is always assumed to be immobile and inert. Hence the lapse of time is a mere accident, the material being indifferent to it. The world is simply fabricated of a succession of instantaneous immobile configurations of matter (i.e., a succession of equilibria), each instant bearing no inherent reference to any other instant of time. How then, is it possible to link cause and effect? How are we justified to infer from observations, the great "laws of nature"? This is essentially the problem of induction raised by Hume. The problem is created because we have mistaken the abstraction for reality – a case of the fallacy of misplaced concreteness."

The further tentative way of Mae-Wan Ho to explain time in the framework of quantum mechanics is exercise without concreteness.

Referring to time as the fourth dimension of space or assuming a space–time structure is definitely wrong. Space and time are two different philosophical categories.

2.4. Einstein's twins

Prigogine maintains that Bohr was right in the famous discussion with Einstein, but when quantum theory was formulated, dynamic instability and chaos were not in the perspectives of normal physics. These concepts are now essential for the self-consistency of quantum physics. The subjective aspects of quantum mechanics have been eliminated. Poincaré's divergence is a mathematical fact independent of the observer. This leads to a new formulation of quantum theory and forces us to accept a view of nature that includes instability and dissipation.

In the work of Prigogine, the foundations of evolutionary physics are laid. The scientific schizophrenia that separated evolutionary biology and mechanics is overcome and physics and biology find a meeting point. We now have glimpses of the potential of new instruments of physical chemistry for tackling the complexity of living systems and evolving ecosystems. Ecodynamic models begin to give the first results in complex fields such as sustainable towns, integrated agro-energy systems, biomass and aquatic ecosystems. Physical chemistry comes down from its pedestal of pure theory, abandoning unreal molecular monads for the real world. Without loss of rigour or scientific character, it leaves its ivory tower of useless purity and comes to grips with nature, society, the *bios* and the *oikos*, the complexity of the true problems of humans and the planet.

According to Prigogine[8] there are only two choices: either everything was determined at the moment of the Big Bang (including this book) or the universe evolves and novelties appear during its evolution, in which case the laws of nature cannot be deterministic but have to be expressed in terms of probabilities. He recalls Einstein's claim that the irreversibility of time was an illusion. Prigogine attempts to explain Einstein's attachment to this idea on the basis that Einstein was living in a very difficult period and that science was a way of escaping the turmoil of everyday life. Prigogine asks if this view is still valid today. Is today's science a way of escaping from cities, or should science tackle the problem of pollution and improve city life?

Time, in Einstein's view, is mechanical and reversible, a mere mental category. On this basis Einstein developed the famous twin paradox. According to Einstein's theory, the relationship between time and the velocity of light can be used to demonstrate that a twin, travelling at a speed close to the speed of light, ages less than the other twin. This is based on the following well-known equation:

$$t = t_0 / \sqrt{(1 - v^2/c^2)} \qquad (1)$$

where t is time, v velocity of the twin and c the velocity of light.

In mechanics, velocity is the first derivative of distance with respect to time (ds/dt). This algorithm is useful for studying the trajectories of the planets, the motion of cars, trains and so on. It is also useful for comparing the velocity of a living system relative to a frame of reference, for example, a runner on a track. However, in our opinion, it is not scientifically correct to compare the life span of a living organism (one of the twins) with the life span of another living organism (the second twin). These are two complex

---

[8] I. Prigogine, personal communication.

systems, both far from thermodynamic equilibrium, both with dissipative properties and capable of self-organization; i.e. two evolving systems.

In this context, space and time are categories belonging to different logical types, which should not be confused. By nature, time is evolutionary and irreversible, whereas space is conservative and reversible. A reversible quantity cannot be differentiated with respect to an irreversible one. It is not possible to compare evolving quantities, such as the life span of the twins, in the framework of reversible mechanics.

Life, like all living properties, is irreversible and follows biological, evolutionary paths (παντα ρει). Since the twins' lives also follow such paths, no paradox exists.

Moreover, in order to differentiate, it is necessary to know the right and left limits: in our case the right limit (the future) is unknown and the derivative does not exist.

The problem of the irreversibility of time lies with the parametrized idea of the concept of time: time is usually considered an abstraction, a Kantian category, not a property of matter, not a tangible existing fact. However, in nature, systems never exactly repeat themselves after an interval of time.

In modern mathematics and physics, awareness of this concept is spreading with the theory of deterministic chaos, which admits the existence of strange attractors in natural complex systems. We can think of time and space as being in some ways conceptually independent of each other.

2.5. Poincaré

Prigogine used Poincaré's theorem of 1889 to make an alternative formulation of dynamics described in terms of probabilities. This description introduces the arrow of time into physics. Prigogine reflects that the cosmological arrow of time and the biological arrow of time are everyday terms. With this new description, he and his team hoped to have identified the common root of all these arrows of time. They hoped to be able to apply it in a concrete way to the problems of ecology.

Systems and evolutionary approaches, or to use a term we coined at the University of Siena, ecodynamic models, are required to tackle the major problems of ecology (greenhouse effect, the hole in the ozone layer, acid rain, eutrophication, etc.). Ecodynamic models are necessarily based on the thermodynamics of Prigogine, on the science of dissipative chaos and far-from-equilibrium processes.

Returning to the theorem of Poincaré, Prigogine recalls that Poincaré asked himself whether the physical universe was isomorphic to a system of non-interacting units. Energy (the Hamiltonian, H) is generally written as the sum of two terms: the kinetic energy of the units involved and the potential energy of their interactions. Poincaré asked whether the interactions could be eliminated. This is a very important question. If the answer is "yes", then there may be no coherence in the universe. It was therefore lucky that he proved that interactions cannot generally be eliminated, because of resonances between the various units.

The Brussels school worked for years on these problems, identifying a class of dynamic systems known as LPSs for which it is possible to eliminate Poincaré divergence and "integrate" a class of non-integrable Poincaré systems. A LPSs is a system with a continuous spectrum, characterized by interactions involving integrations over resonances. These LPSs are not integrable in the usual sense because of Poincaré resonance, but can be integrated by new methods, eliminating any Poincaré divergence.

Since we are dealing with chaotic systems, chance plays an increasing role and time symmetry is broken. This puts irreversibility in the core of ecodynamics. Prigogine observed that the usual time paradox is somehow inverted. Normally one tried to deduce the arrow of time from dynamics based on time-reversible equations. Now we generalize the dynamics to include irreversibility. Poincaré divergences are eliminated by appropriate time ordering of dynamical states. Two classes of LPSs can be distinguished. In the first class, the canonical equations of motion in classical mechanics or the Hilbert space in quantum mechanics are time ordered. The results are trajectories of wave functions that are no longer symmetrical with respect to time. This is the simplest example unifying dynamics and thermodynamics. In the second class, Poincaré divergence can be eliminated by time ordering of the statistical description, to obtain equations for the evolution of the probability distribution. The trajectories behave in a chaotic manner, whereas the probability distribution satisfies a simple diffusion-type equation. Poincaré divergence leads to trajectories or wave functions that are both irreversible and stochastic.

In classical mechanics, we deal with numbers, whereas in quantum mechanics we deal with operators. In quantum mechanics, the Hamiltonian H is associated with eigenvalues of eigenfunctions. The set of eigenvalues of the eigenfunctions is called a spectral representation. This is a key problem of quantum mechanics and can only be solved in a few simple situations; usually perturbation methods have to be used. For non-integrable Poincaré systems, expansion of the eigenfunctions and eigenvalues into powers of the coupling constant leads to the so-called Poincaré catastrophe, due to the divergence associated with small denominators. Conventional perturbation methods do not solve the problem. In order to make non-integrable Poincaré systems "integrable", Prigogine introduces "natural time ordering" of dynamic states. This means that first we have the unstable atomic state and then emission of radiation. This corresponds to Bohr's picture in which the radiation emitted by the atom is a "delayed" wave.

To introduce time ordering, the Brussels school used the statistical description of Boltzmann. More than a century ago, Boltzmann obviously did not know the theory of chaos, and could not have imagined that he was studying non-integrable Poincaré systems. Along with Maxwell, he invested his hopes in ergodic theory, which is very useful for understanding equilibrium, but is no good for dynamic processes. Prigogine refers to sets introduced by J.W. Gibbs in the early period of statistical mechanics. Gibbs ensembles are obtained by considering relations between particles: the particles collide and the collisions give rise to binary and then to higher order relations in time.

Prigogine illustrates this question with the example of the glass of water. The glass contains the arrow of time in the sense that new relationships are created involving more and more particles of water. "Old" water can be distinguished from "new" water. This flow of correlations orientated in time breaks the symmetry of time involved in the classical description.

Let us imagine a transition from state A at time zero, characterized by no correlations, to state B at time *t*, characterized by many correlations. The transition from A to B involves different physical processes than the inverse transition from B to A.

Time ordering of correlations must be introduced into dynamics to avoid the Poincaré catastrophe and so that dynamics can be described in terms of the temporal evolution of the correlations. This involves a genuine change in gestalt with respect to classical dynamics: we are no longer studying positions and momentum of particles in time but following the evolution of the relations between particles[9].

The ensemble theory of Gibbs has an equation for time evolution of the density matrix, which is formally similar to Schrödinger's equation:

$$i\frac{\partial \rho}{\partial t} = L\rho \qquad (2)$$

where L is the Liouville operator, which can be expressed in terms of the Hamiltonian H in classical and statistical mechanics.

---

[9] I. Prigogine, *Non Equilibrium Statistical Mechanics*, Wiley Interscience, New York, 1962.

A collision is a complex process in which particles approach each other, exchange energy by resonance and continue in different directions. A collision can be imagined as a sequence of states related by resonance. Prigogine writes that in a Hamiltonian system, a collision is not an instantaneous point event, but is extended in space and time. The spectrum of the Liouville operator is determined essentially by collision dynamics. This deviates radically from the usual methods of dynamics that hold for integrable systems in which evolution can be resolved into a series of instantaneous events in space and time. This is why the dynamics of LPSs can only be formulated at a statistical level; it cannot be reduced to trajectories as in classical dynamics or to wave functions as in quantum mechanics.

Integration of non-integrable Poincaré systems thus becomes the basis of a new evolutionary physics that includes irreversibility and broken time symmetry. The time paradox is eliminated: in the old situation, there was no bridge between the reversible microscopic level and the irreversible macroscopic level. Now we have a new unstable microscopic level with broken time symmetry, from which a dissipative macroscopic level emerges.

Although Poincaré's divergences lead to a new formulation of dynamics, the approach is still reductionist (based on elementary particles).

The real change of paradigm has to be based on Blaise Pascal's view and on new evolutionary ideas about uncertainty, probability, goal functions and configurations of processes.

3
THIRD STEP

THERMODYNAMIC UNCERTAINTY

The voice that spoke was certainly that of our master.
He knows how to bring together traces dispersed here and there.
... He observed the stars and traced their position and orbits in the sand;
he watched the sea of air, never tiring of its clarity, movements, clouds,
lights.
... He considered men and animals, seated on the shore, he looked for
shells.
... He linked distant things. Now the stars were men, now men were stars,
stones were animals, clouds were plants; he played with forces and
phenomena;
he knew how and where to find and evoke them.

Novalis[1]

3.1. Heisenberg

"At the instant when position is determined, the electron undergoes a discontinuous change in momentum. This change is the greater the smaller the wavelength of the light employed – that is, the more exact the determination of the position.

---

[1] Novalis (F.L. von Hardenberg), *The Novices of Sais*. Original: *Die Lehrlinge zu Sais*, 1789. The work is an unfinished symbolic novel of the German Romanticism: Friedrich von Hardenberg, alias Novalis, died of tuberculosis at 29 years of age. Sais was a major religious and cultural centre near Alexandria in ancient Egypt.

Thus, the more precisely the position is determined, the less precisely the momentum is known, and conversely." (Heisenberg)[2]

According to the laws governing the Compton effect, p_1 and q_1 are related by:

$$p_1 q_1 \approx h \tag{3}$$

$$E_1 t_1 \approx h. \tag{4}$$

Equation (4) is equivalent to eqn (3) and shows that precise determination of energy can only be had at the cost of a corresponding uncertainty in time.

Another relation can be derived from the uncertainty between position and momentum. Let v and E be the velocity and energy corresponding to momentum p_x, then

$$v \Delta p_x \frac{\Delta x}{v} \geq h \tag{5}$$

$$\Delta E \Delta t \geq h \tag{6}$$

where ΔE is the uncertainty of energy corresponding to the uncertainty of momentum Δp_x, and Δt is the uncertainty in time within which the particle (or the wave packet) passes over a fixed point on the x-axis[3]. *Irreversibility of time is not considered, since in the quantum mechanical paradigm, time is assumed to be reversible.*

3.2. Spin relaxation

Spin relaxation is possible because the spin system is coupled to the thermal motion of the "lattice", be it gas, liquid or solid. The fundamental point is that the lattice is at thermal equilibrium; this means that the probabilities of spontaneous upward and downward spin transitions are not equal, as they were for rf-induced transitions.

---

[2] W. Heisenberg, Über den anschaulichen Inhalt der quantentheoretischen Kinematik und Mechanik, *Zeitschrift für Physik*, **43**, 172–198, 1927; English translation in Wheeler and Zurek, 62–84, 1983.
[3] P. Fong, *Elementary Quantum Mechanics*, Addison-Wesley Publishing Company, Massachusetts, USA, 1962.

Denoting the upward and downward relaxation probabilities by $W_{\alpha\beta}$ and $W_{\beta\alpha}$ (with $W_{\alpha\beta} \neq W_{\beta\alpha}$), the rate of change of N_α is given by

$$\frac{dN_\alpha}{dt} = N_\beta W_{\beta\alpha} - N_\alpha W_{\alpha\beta} \tag{7}$$

At thermal equilibrium $dN_\alpha / dt = 0$, and denoting the equilibrium population by $N_{0\alpha}$ and $N_{0\beta}$ we see that

$$\frac{N_{0\beta}}{N_{0\alpha}} = \frac{W_{\alpha\beta}}{W_{\beta\alpha}}. \tag{8}$$

The populations follow from Boltzmann's law and so the ratio of the two transition probabilities must also be equal to exp($-\Delta E/kT$). Expressing N_α and N_β in terms of N and n ($n = N_\alpha - N_\beta$) we obtain

$$\frac{dn}{dt} = -n(W_{\beta\alpha} + W_{\alpha\beta}) + N(W_{\beta\alpha} - W_{\alpha\beta}). \tag{9}$$

This may be rewritten as

$$\frac{dn}{dt} = -\frac{(n - n_0)}{T_1} \tag{10}$$

in which n_0, the population difference at thermal equilibrium, is equal to

$$n_0 = N\left[\frac{W_{\beta\alpha} - W_{\alpha\beta}}{W_{\beta\alpha} + W_{\alpha\beta}}\right] \tag{11}$$

and $1/T_1$ is expressed by

$$\frac{1}{T_1} = W_{\alpha\beta} + W_{\beta\alpha}. \tag{12}$$

Therefore T_1 has the dimensions of time and is called the "spin-lattice relaxation time". It is a measure of the time taken for energy to be transferred to other degrees of freedom, that is, for the spin system to approach thermal equilibrium; large values of T_1 (minutes or even hours for some nuclei) indicate very slow relaxation (Carrington and McLachlan[4]).

However spin-lattice relaxation is by no means the only mechanism of line broadening. Many other processes that occur in both solids and liquids have the effect of varying the *relative energies* of the spin levels, rather than their lifetimes. Such processes are characterized by a relaxation time T_2, often called the spin–spin relaxation time but more satisfactorily, the transverse relaxation time[4] (see Fig. 1).

These remarks may seem to suggest that there is little connection between T_1 and T_2. On the contrary, these quantities are closely related because both modes of relaxation depend on the same interactions between the spins and their surroundings. Thus those interactions that lead to a finite lifetime for the spin states may also modulate the energy levels.

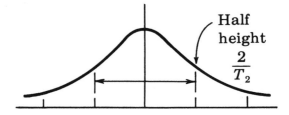

Figure 1: NMR line shape: Lorentz line.

It is now possible to say something about the width and shape of the resonance absorption line, which certainly cannot be represented by a Dirac δ function.

It is clear that, because of spin relaxation, spin states clearly have a finite lifetime.

The resulting line broadening can be estimated from the *uncertainty relation*

$$\Delta v \, \Delta t \approx 1 \qquad (13)$$

[4] A. Carrington and A.D. McLachlan, *Introduction to Magnetic Resonance*, Harper & Row, New York, USA, 1967.

and we find that the line width due to spin-lattice relaxation is of the order of $1/T_1$; the broadening is due to the finite lifetime a spin is in a given energy state.

3.3. The life span of stars

The uncertainty equations (3), (4) and the derived equations (6) and (13) indicate the complex relation between the observer and the experiment. The former deals with position and momentum, the latter with energy and relaxation time. Both sets of equations assume the reversibility of time and are valid at a given instant: the momentum is related to the derivative of space with respect to time and the relaxation time is related to the lifetime of the elementary particle in the excited state.

Both sets of equations are valid in the quantum mechanical paradigm and deal with conservative quantities (mass, energy) and not with living systems or evolutionary quantities.

Space and time are categories belonging to different logical types, which should not be confused. By nature, time is evolutionary and irreversible, whereas the space is conservative and reversible. A reversible quantity cannot be differentiated with respect to an irreversible one, as we saw in the Second Step.

It is not possible to compare evolving quantities, such as the life span of Einstein's twins, in the framework of reversible mechanics.

Rather, it is possible to link these concepts with the generalized uncertainty associated with the presence in the Universe of both conservative (space, mass) and evolutionary quantities (time, life span).

Whereas irreversibility of time is not taken into account by Werner Heisenberg, since in the quantum mechanical paradigm time is assumed to be reversible (the corresponding equations deal with conservative quantities: mass, energy).

In dealing with evolutionary (living) systems, we may introduce a third concept: that of *Thermodynamic Uncertainty* related to the intrinsic irreversible character of time.

Let us say that a thermodynamic uncertainty arises from the experimental existence of the arrow of time and from the experimental evidence that, during the measurements, time goes by.

Since time flows during the interval of an experiment (measurement), conservative quantities (energy and/or position) may also change leading to further uncertainty.

Astrophysicists have recently discovered that the mass of a star is related to the star's life span; the greater the mass, the shorter the life span. This too may be related to the uncertainty principle.

It seems that there is a sort of uncertainty relation between space and time, space being related to mass and energy, which are conservative quantities.

3.4. Cogito ergo sum (Descartes)

This paragraph has been contributed by Franco Borghero (Padua), doctor in physics and psychoanalyst. It should be read in relation to the Tenth Step and to the semantic implications discussed in the Ninth Step.

The use made of logic and mathematics is irreplaceable in many fields of knowledge, but not all aspects of life can be investigated with these powerful tools. Incorrect or inappropriate use may lead to gross errors.

What follows is a demonstration that if one remains on the plane of logic, one may run into inextricable paradoxes. The demonstration is partly borrowed from the 11th seminar held by Jacques Lacan[5].

According to formal logic, if *A implies B* then *non B implies non A* (e.g. if 3 is less than 4, then –4 is less than –3, or if *I am Italian implies that I am European*, then *I am not European implies that I am not Italian*).

Formal logic is a powerful tool that enables us to establish solid theses from acceptable hypotheses. However, not all expressions of thought can be analyzed by this method. We show that some lead to conundrums.

Crystals obey rigorous laws and their forms are clear and precise: desert roses, water crystals and amethyst druses are astoundingly beautiful. This beauty can be translated and read with the mathematical laws governing crystal formation. We may feel the same wonder for a *rose*, the beauty of which originates in life. In this case the beauty cannot be expressed or translated into logic or mathematics; other parameters must be used, for example sensitivity.

When we have to do with life or the spirit, the tools of formal logic do not help us to understand them. Those who try end up in a blind alley. Let me illustrate this with the case of the Cartesian *cogito ergo sum*. Analyzed by formal logic, we obtain nonsense; translating it into mathematics, as did Jacques Lacan, we obtain an incomprehensible result.

---

[5] J. Lacan, French psychoanalyst (Paris, 1901–1981). He discussed the Cartesian *cogito ergo sum* in his 9th seminar *L'identificazione*, unpublished.

If *cogito ergo sum* is true, then according to formal logic *non sum ergo non cogito* must also be true. The latter is of course nonsense, because if one does not exist how one say that one does not think? Moreover, if a non-being can say that it does not think, how can it say this if saying involves thinking?

Clearly, to say that "one is" requires thought, but this statement does not mean that being is deduced by thinking. It only means that in order to be able to say that "one is" it is necessary to use thought. In other words, awareness of being does not depend on thinking; or, less cryptically, the feeling of being, though it can more or less be expressed in words, has no relation to thought, unless we define pain and happiness as thought, along with all emotions that are difficult to define and that only metaphor can probably clarify (which only poets do well). But the very need to use metaphor tells us that the essence of the sentiment is not thought, because in order to be profoundly understood, thought needs to be translated into words that do not refer directly to a thing or concept.

Psychoanalyst Jacques Lacan was interested in philosophy and proposed an intriguing reading of the famous maxim of Descartes. He reasons, "I think and therefore I am. What am I if not one who thinks? But since I am one who thinks, I am thinking. That is: I think and therefore I am one who thinks ..." and so on. Thus he translates the iteration of the famous phrase as "I think, therefore I am one who thinks, therefore I am one who thinks, therefore etc." which enables him to work towards a limit. Thus he writes:

$$I\ think\ therefore\ I\ am + \cfrac{I\ think}{I\ am + \cfrac{I\ think}{I\ am + \cfrac{I\ think}{I\ am + \cfrac{I\ think}{I\ am + etc.}}}}$$

iterated to infinity.

Written in this way, Lacan translates it into a new formula which is finally mathematical.

Thinking is an undeniably certain and significant fact: for simplicity it can be signified by 1 (or P). Lacan goes on to say that if the statute of being were the same as thinking, we could indicate being with 1 (or E). Lacan then proposes this new scheme which is a mathematical series:

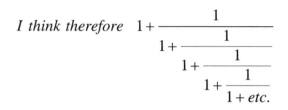

$$I\ think\ therefore\quad 1+\cfrac{1}{1+\cfrac{1}{1+\cfrac{1}{1+\cfrac{1}{1+etc.}}}}$$

Calculating the limit, the series turns out to converge towards the finite value: $(1+\sqrt{5})/2$, which would come to be (*sic!*) the ultimate meaning of a thinking being. It is now easy to realise that the starting series is an unacceptable symbolisation of *cogito ergo sum*. Lacan realised this incongruence (i.e. the fact that being cannot be written 1 or E) and proposed symbolising it with something that does not exist, namely something completely imaginary not containing anything real. Mathematics gives him a good tool: i (the number *one* in the imaginary numbers, which is a number that does not exist, $i=\sqrt{-1}$. The arithmetic of i is: $i^2=-1$; $i^3=-\sqrt{-1}=-i$; $i^4=1$; $i^5=\sqrt{-1}=i$).

The series proposed by Lacan (where *I think* is symbolised by 1 and *I am* by i becomes:

$$I\ think\ therefore\quad i+\cfrac{1}{i+\cfrac{1}{i+\cfrac{1}{i+\cfrac{1}{i+etc.}}}}$$

Note that Lacan does not consider the premise *I think therefore*. He then calculated the value of this new series which no longer converges or diverges but gives three values: $i+1$; $(i+1)/2$; 1. If we attempt to read them (preceded by the premise *I think therefore*):

the first is: $i+1 = $ (*I think therefore*) I am one who thinks;

the second is: $(i+1)/2 = $ (*I think therefore*) I am one who thinks, divided by two;

the third is: $1 = $ (*I think therefore*) I think.

Lacan concludes: at the first level, the subject who *thinks* about himself *is one who thinks*. At the second level, the subject who *thinks* about himself *is one who thinks divided in two.* At the third level, the subject who *thinks* about himself *thinks*.

This brilliant conclusion of Lacan's likewise does not seem to tell us much about the relation between thinking and being. Lacan was not satisfied and proposed a new formulation of *cogito ergo sum*: *where I think, I am not and where I am, I don't think*. One may approve this in principle, but it clearly stretches the Cartesian maxim.

Here I would like to make a purely formal objection that provides material for thought. If we accept the symbols used by Lacan *(I think* = 1, *I am* = *i*, *I think therefore I am* = 1/*i*) and avoiding the iterations, and we examine more closely the transcription of the statement *I think therefore I am* with 1/*i*, we see a contradiction. Indeed 1/*i* = −*i*, the translation of which is: *I think therefore I am is equivalent to I am not*, which is clearly nonsense.

Nevertheless, we continue and introduce the premise that Lacan neglected into the series:

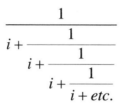

the limit of which gives us three results that repeat to infinity but which are fundamentally different, namely: −*i;* ∞; 0; etc. These results provide a different reading;

the first term is: −*i* = *I am not;*

the second: ∞ = *infinity;*

the third: 0 = *null.*

In other words, to think of oneself as thinking is equivalent to losing oneself between non-being, infinity and nothing.

Although this is a more interesting reading, it tells us nothing of the possibility of deducing being from thinking. Instead, it seems that if one thinks, one is not. In conclusion, being cannot be deduced from thinking, irrespective of whether it is investigated by formal logic or by mathematics. I therefore consider that the only tool we can use to try to understand the maxim is another type of logic.

Just as the beauty of a rose cannot be understood using tools such as logic and mathematics, *cogito ergo sum* cannot be understood through formal logic or mathematics, but only through a new tool that considers the various planes of the words *to think* and *to be*.

Thought is an epiphany of being which can only be glimpsed by thought: from my window, I can only see a red house and a strip of sky but I know that there is a city behind that house and beyond the sky there are stars.

The same is true for probability and uncertainty: as with abstract concepts for studying a game of cards, matter and conservative quantities can be used, but if I want to use them to study a chameleon (see Tenth Step) or any other living system, they lead to false conclusions. There is no way they can explain evolutionary systems, simply because time affects events (and hence the succession of probabilities) and the intrinsic uncertainty of measurements, which are made while time passes. Experiments are therefore valid for that instant or for systems that do not change considerably in time.

Moreover, it is not possible to measure lifespan, evolution and internal time of living systems only with a machine, a mechanical tool.

There is a real watershed between non-living systems and living systems, between classical physical chemistry and evolutionary physics (see Fourth Step).

4
FOURTH STEP

THE DEATH OF THE DEER

Colours and Entropy

Nature sometimes
to alchemy turns
inventing colours
full of magic.

It mixes manganese
with the mountains
to make stones
a kindly faded pink
like trachite
in the Bosa caves.

By salts of nickel
and mines of cobalt
it turns the grey basalt
into azure.

When the fog comes
it chooses rare earths
and with indigo paints
the mantle of the sea.

It uses of the elements
the rare metals
and before the merops
comes to rest,
behind her back,

it paints a dress
of silver and gold
woven centuries ago.

It uses red of iron
for blood and life
but if metals
from its fingers slip
entropy mixes them
and makes them white.

This always happens
when Nature gets tired.

Enzo Tiezzi[1]

4.1. Classical thermodynamics

The steps towards evolutionary physics are also based on some fundamental statements of classical thermodynamics. This book is a sequel to two other volumes[2,3] forging the way to evolutionary physics. The first paragraphs of the present chapter quote and elaborate on passages from "The End of Time"[2].

Die Energie der Welt ist Konstant (the energy in the world is constant; R.Clausius 1865). The First Law of Thermodynamics says that the total energy existing in the universe in its various forms is invariant; it can only be transformed from one form to another, but the total of the various forms remains constant: this first law is also known as the *conservation of energy*.

Julius Robert Mayer was the philosophical father of the first law. He sailed from Rotterdam in February 1840 on the "Java" as ship's doctor. After a long voyage the ship berthed at Surabaya. Here some of the sailors required medical care and Mayer noticed the bright red colour of their blood. A local doctor told him that the colour was typical of blood in the tropics because the

---

[1] E. Tiezzi, *La più bella storia del mondo*, Marcos y Marcos, Milan, 1998.
[2] E. Tiezzi, *The End of Time*, WIT Press, Southampton, UK, 2003.
[3] E. Tiezzi, *The Essence of Time*, WIT Press, Southampton, UK, 2003.

quantity of oxygen necessary to maintain the body temperature was less than that required in a colder climate. Mayer began to think: evidently with the same amount of food the body could produce a variable amount of heat, and the body also performed work. If a given quantity of food gave a fixed quantity of energy, the body could use it for heat and/or work; heat and work are interchangeable quantities of the same type but their sum is constant. The beautiful, precise experiments of the Englishman J.P. Joule, performed in his father's beer factory, confirmed and defined the first law.

Since a particle contains energy proportional to its mass (Einstein's law: $E = mc^2$ where E is energy, m mass and c^2 the proportionality constant, the square of the speed of light) and since the conversion of mass to energy can be measured in nuclear reactions, the first law can also be regarded as the conservation of mass and energy.

The first law says that a machine cannot create energy. Lazare Carnot was involved with machines and their energy yield between one reorganization and another of the armies of the Republic. His son, Sadi Carnot, wisely followed in his father's footsteps, studying the energy output of thermal machines and discovered the Second Law of Thermodynamics. This was in apparent conflict with the first law. The Second Law says that energy cannot be freely transformed from one form to another and that thermal energy (heat) can flow freely from a hot to a cold place but never in the opposite direction. The conversion of heat into work is impossible without a temperature difference. The Second Law of Thermodynamics says that a machine cannot transfer heat from a cold body to a hot one without performing work. Whenever work is produced from heat, heat also passes from a hot body to a colder one. Our everyday experience with devices (from motors to electric razors) shows us that work is inevitably accompanied by heating which was not the intention of the machine. There is a tendency in the universe towards the "heat form" of energy. Heat is a "degraded" form of energy because it cannot be totally converted back into work. Only some of the heat can be transformed into work; we cannot freely recover heat from a cold body. For example the ocean is an immense store of heat, it contains an enormous quantity of energy, but we cannot use it gratis. Although it contains much more heat than our bodies, we cannot warm our hands by it because the ocean is a colder source than our hands and heat cannot pass spontaneously from a cold to a warmer body.

At this point we could open a parenthesis on what we might call "Maxwell's demon revisited". In 1871 J.C. Maxwell proposed a paradox that embarrassed physicists for a long time. He imagined a system with gas in two containers A and B at the same temperature, separated by a wall. There was a small aperture in the wall guarded by a demon who separated fast from slow moving molecules (i.e. hot from cold molecules as temperature is a measure of the movement of molecules) putting the first into A and the second into B. In the end there would be a temperature difference in contradiction with the Second Law of Thermodynamics. N. Georgescu-Roegen[4] observes that Maxwell's demon has been exorcised; as any other living creature, the demon must use more energy than he creates by separating hot and cold molecules. He adds that many theories on the unlimited renewability of resources imply a demon with miraculous faculties behind the scenes.

The First Law of Thermodynamics is only concerned with the general energy balance, stating that energy can neither be created nor destroyed. The Second Law is concerned with the use of energy, its availability for work and its natural degeneration into non-utilizable forms. The thing that diminishes in the world is not the amount of energy, but its capacity to perform work. From this point of view, Einstein was right to regard the Second Law as the fundamental law of science; Commoner was right to call it our most profound scientific intuition on the workings of nature; and C.P. Snow was right in saying that to ignore the meaning of the Second Law of Thermodynamics is like admitting to never having read a single work of Shakespeare[5].

4.2. What is entropy?

The spontaneous tendency of energy to degrade and be dissipated in the environment is evident in the phenomena of everyday life. A ball bouncing makes smaller and smaller bounces, dissipating heat. A jug that falls to the ground breaks (dissipation) into many pieces and the inverse process that could be seen running a film of the fall backwards, never happens in nature. Perfume leaves a bottle and dissipates into the room; we never see an empty bottle spontaneously fill. The heat form and dissipation are favoured. The thermodynamic function known as *entropy (S)* is a measure of the degree of energy dissipation. Transformations go spontaneously in the direction of increasing entropy or maximum dissipation. The idea of the passage of time, of the direction of the transformation, is inherent in the concept of entropy.

---

[4] N. Georgescu-Roegen, was a scientist of Rumanian origin who taught economics in the United States. He is known for his application of the principles of thermodynamics to economics. He was the author of *Energy and Economic Myths*, Elmsford, Pergamon Press, New York, 1976.

[5] C.P. Snow, *The Two Cultures*, Cambridge University Press, Cambridge, 1964.

The term was coined by Clausius from $\tau\rho o\pi\eta$ (transformation) and $\varepsilon v\tau\rho o\pi\eta$ (evolution, mutation or even confusion).

With the concept of entropy Clausius reworded the Second Law of Thermodynamics in a wider and more universal framework: Die Entropie der Welt strebt einem Maximum zu (The entropy of the world tends towards a maximum; R. Clausius 1865). Maximum entropy, which corresponds to the state of equilibrium of a system, is a state in which the energy is completely degraded and can no longer produce work.

Entropy is therefore a concept that shows us the direction of events. Commoner notes that sand castles (order) do not appear spontaneously: they can only disappear (disorder); a wooden hut becomes a pile of beams and boards in time: the inverse process does not occur. The direction is thus from order to disorder and entropy indicates this inexorable process, the process that has the maximum probability of occurring. The concepts of disorder and probability are thus linked in the concept of entropy. *Entropy is in fact a measure of disorder and probability.* In order to understand this better, it is useful to describe a model experiment: the mixing of gases.

Suppose we have two gases, one red and one yellow, in two containers separated by a wall. If we remove the wall we see that the two gases mix until there is a uniform distribution: an orange mixture. If they were originally mixed we would not expect to see them spontaneously separate into red and yellow. The "orange" state is that of maximum disorder, the situation of greatest entropy because it was reached spontaneously from a situation of initial order. Entropy is a measure of the degree of disorder of the system. The disordered state occurred because it had the highest statistical probability. The probability of there being 13 hearts in a hand of bridge is 1 in 635,013,559,600. In other words, such a hand is almost impossible; a mixed hand, with a few cards of each suit, is the most probable. The law of increasing entropy is therefore a law of probability, of a statistical trend towards disorder. The most likely state is reached, namely the state of greatest entropy or disorder. When the gases mix, the most probable phenomenon occurs: degeneration into disorder.

The universality of the law of entropy increase was stressed by Clausius in the sense that energy is degraded from one end of the universe to the other and that it becomes less and less available in time, until "Wärmetode", or the "thermal death" of the universe.

Evolution towards the thermal death of the universe is the subject of much discussion. In order to extend the theory from the planetary to the cosmic context it is necessary to introduce unknown effects such as gravitation. Current astrophysics suggests an expanding universe that originated in a great primordial explosion (Big Bang) from a low-entropy state, but the limits of theoretical thermodynamic models cannot confirm this or provide evidence.

The study of entropy continues: this fundamental concept has been applied to linguistics, the codification of language and to music and information theory. Thermodynamics has taught us two fascinating lessons: that energy cannot be created or destroyed but is conserved, and that entropy (S) is always increasing, chiming the hours of the cosmic clock and reminding us that for man and energy-matter alike, time exists and the future is distinct from the past by virtue of a higher value of S.

Although thermodynamics reigns throughout science from mechanics to nature, the natural biological laws of evolution seem to contradict it. Biological systems apparently violate the Second Law, they show extremely ordered structures and evolve in a direction of increasing order or less entropy. The contradiction is really only one of appearance. The entropy balance must be total and must include both biological organisms and the environment with which they exchange energy and matter. Thus biological organisms develop and live by virtue of the increased entropy occasioned by their metabolisms in the surrounding environment. The total change in entropy is positive, the entropy of the universe increases and the Second Law is not violated.

If bacteria are allowed to reproduce in glucose solution, part of the sugar can be seen to cause a decrease in entropy, being transformed into cell components. The rest is transformed into carbon dioxide and water, leading to an overall increase in entropy.

It is necessary to distinguish between *isolated systems* (which cannot exchange energy or matter with the outside world), *closed systems* (which can exchange energy but not matter, e.g. our planet) and *open systems* (which can exchange both energy and matter). Cities and biological organisms are examples of open systems. For such systems we must sum the negative entropy produced inside the system with the positive entropy created in the environment. We then see that if "sometimes disorder degenerates into order", it is only a façade, the appearance of order at the price of even greater disorder in the surrounding environment. Living systems therefore need a continuous flux of negative entropy from the universe, to which they return an even larger amount of positive entropy. Prigogine called these open systems "dissipative structures". The flow of energy causes fluctuations in dissipative structures which reorganize, attaining higher levels of complexity. As such they require an even higher energy input and are even more vulnerable to fluctuations. They reorganize again in continuous biological evolution towards complexity and higher energy needs. All this occurs at the cost of increased entropy of the environment. There is no need to invoke chance (as did Monod[6]) to explain biological evolution[7]. The thermodynamics of open systems, dissipative structures, mean that the most likely event always occurs.

4.3. Photosynthesis and entropy

As long ago as 1886, Ludwig Boltzmann[8], one of the fathers of modern physical chemistry, was concerned with the relation between energy and matter. According to Boltzmann, the struggle for life is not a struggle for basic elements or energy but for the entropy (negative) available in the transfer from the hot Sun to the cold Earth. Utilizing this transfer to a maximum, plants force solar energy to perform chemical reactions before it reaches the thermal level of the Earth's surface.

To live and reproduce, plants and animals need a continuous flow of energy. The energy of the biosphere, which originates in the luminous energy of the Sun, is captured by plants and passes from one living form to another along the food chain. The energy captured by chlorophyll is stored in carbohydrates (molecules rich in energy) by means of photosynthesis, a term that means "to make things with light". This radiant pathway, that provides us with great quantities of food, fibres, and energy – all of solar origin – has existed for about four billion years, a long time if we think that hominids appeared on the earth only three million years ago and that known history covers only few thousand years. Our ancestor, the blue alga, began to photosynthesize, playing a fundamental role in biological evolution.

The organization of living beings in mature ecosystems slows the dispersal of energy fixed by plants to a minimum, enabling them to use it completely for their own complex mechanisms of regulation. This is made possible by

---

[6] J. Monod, *Le hasard et la nécessité*, Le Seuil, Paris, 1970.

[7] On this point Prigogine writes in *La Nouvelle Alliance*: "Life, regarded as the result of <improbable> initial conditions, is in this sense compatible with the laws of physics (the initial conditions are arbitrary), but does not follow from the laws of physics (which do not set the initial conditions). This is the view of Monod, for example, in his book[6]. Furthermore life, from this point of view, looks like a continual struggle by an army of Maxwell devils against the laws of physics, to maintain the highly improbable conditions which allow it to exist. Our point of view is completely different in that vital processes, far from being outside nature, follow the laws of physics, though in specific nonlinear interactions and in conditions far from equilibrium. These aspects can in fact provide the flow of energy and material necessary to build and maintain functional and structural order".

[8] G.Giacometti, personal communication, from R. Huber, *Angewante Chemie*, **28**, 848, 1989.

large "reservoirs" of energy (biomasses) and by the diversification of living species. The stability of natural ecosystems, however, means that the final energy yield is zero, except for a relatively small quantity of biomass that is buried underground to form fossils for the future.

Photosynthesis counteracts entropic degradation insofar as it orders disordered matter: the plant takes up disordered material (low-energy molecules of water and carbon dioxide in disorderly agitation) and orders it using solar energy. It organizes the material by building it into complex structures. By capturing solar energy and decreasing the entropy of the planet, photosynthesis paved the way for evolution.

Living systems on the Earth need a continuous flow of negative entropy (that is, energy from outside) and this flow consists of the very solar energy captured by photosynthesis. This input of solar energy is what fuels the carbon cycle.

The history of life on the Earth can be viewed as the history of photosynthesis and the history of evolution: this singular planet learned to capture solar energy and feed on the negative entropy of the universe for the creation of complex structures (living organisms).

The Sun is an enormous energy source that offers the Earth the possibility of large quantities of negative entropy (organization, life), allowing a global balance that does not contradict the Second Law of Thermodynamics. Every year the Sun sends the Earth 5.6×10^{24} J of energy, more than 10,000 times more energy than mankind consumes in a year. It is as if the Sun sends us 260 tons when we consume only 15 kg.

The decrease in entropy (negentropy) in the biosphere depends on its capacity to capture energy from the Sun and to retransmit it to space in the form of infrared radiation (positive entropy). If retransmission is prevented, in other words, if the planet were shrouded in an adiabatic membrane (greenhouse effect), all living processes would cease very quickly and the system would decay towards the equilibrium state, that is, towards thermal death. A sink is just as necessary for life as a source.

Morowitz[9] points out that all biological processes depend on the absorption of solar photons and the transfer of heat to celestial sinks. The Sun would not be a negentropy source if there were not a sink for the flow of thermal energy. The surface of the Earth is at a constant total energy, re-emitting as much energy as it absorbs. The subtle difference is that it is not energy *per se* that makes life continue but the flow of energy through the

---

[9] H.J. Morowitz, *Foundations of Bioenergetics*, Academic Press, New York, 1978.

system. The biosphere can be defined as that part of the Earth's surface which is ordered by a flow of energy through the process of photosynthesis.

All biological processes take place because they are fuelled by solar energy. Morowitz[9] points out that it is this tension between photosynthetic construction and thermal degradation that sustains the global operation of the biosphere and the great ecological cycles.

This entropic behaviour marks the difference between living systems and dead things.

4.4. The entropy paradox: energy versus entropy

The role of entropy is fundamental in far-from-equilibrium thermodynamics. We may say that entropy exists *per se*, is a non-conservative function and is related to evolution. This conclusion is a big step because it overcomes the old dilemma of whether entropy was the shadow of energy or *vice versa*, and does not reduce the ingenious invention of entropy to an energy dogma. The First Principle formulates the concept of energy in a conservation framework; the Second formulates that of entropy in an evolutionary framework. This is where evolutionary biology and mechanics meet. Schrödinger's introduction of the concept of negentropy[10] was an inspired one: a living system absorbs negentropy from the external environment, structuring itself and evolving on the basis of this interaction. In other words, energy and entropy can be related as is done in classical thermodynamics and statistical thermodynamics but from the point of view of time the two concepts are irreducible and different. In an evolutionary *gestalt*, entropy has an "extra gear" that is the key necessary for studying living systems and ecology. It is important to study flows of energy and matter, quantities that are intrinsically conservative; it is also important to study entropy flows, an intrinsically evolutionary and non-conservative quantity. The appearance of a term of entropy production, or "source term" as Aoki[11] calls it, is the watershed dividing the evolutionary world from the special case of conservative energy and mass. *But if energy and mass are intrinsically conservative and entropy is intrinsically evolutionary, how can entropy be calculated on the basis of energy and mass*

---

[10] E. Schrödinger, *What is life? The Physical Aspects of Living Cells*, Cambridge University Press, Cambridge, 1944.

[11] I. Aoki, Entropy laws in ecological networks at steady state, *Ecological Modelling*, **42**, 289–303, 1988; Monthly variations of entropy production in lake Biwa, *Ecological Modelling*, **51**, 227–232, 1990.

quantities (entropy paradox)? This question is still unanswered[12] and all we can do is to note that the ecodynamic viewpoint is different from that of classical physics and classical ecology.

Let us observe the different relations of energy and entropy versus information.

An energy flow can lead to destruction (increase in entropy, e.g. a cannon ball) or organization (decrease in entropy, e.g. photosynthesis). The same quantity of energy can destroy a wall or kill a man; obviously the loss of information and negentropy is much greater in the second case. Energy and information are never equivalent.

The classical example of the mixing of gases in an isolated system shows us that there can be an increase in entropy without energy input from outside. The point is that E and S are both state functions, but energy is intrinsically reversible whereas entropy is not.

Entropy has the broken time symmetry of which Prigogine speaks. In other words, entropy has an energy term plus a time term that energy does not have.

Entropy has an intrinsic temporal parameter. Energy obeys spatial and material constraints; entropy obeys spatial, material and temporal constraints.

If history and the succession of events are of scientific relevance, the concept of state function should be revised at a higher level of complexity. The singularity of an event also becomes of particular importance: if a certain quantity of energy is spent to kill a caterpillar, we lose the information embodied in the caterpillar. However, were this the last caterpillar, we should lose its unique genetic information forever. The last caterpillar is different from the *n*th caterpillar.

Stories take place in a setting, the details of which are not irrelevant to the story. What happens in the biosphere, the story of life, depends on the constraints of the biosphere itself. Hence it is important to have global models of the biosphere in terms of space, time, matter, energy, entropy, information and their respective relations.

Finally, if we consider the evolutionary transition from anaerobic to aerobic living systems, the ratio of energy to stored information is clearly different. The information that led to the evolution and organization of the two types of system is not proportional to the flow of energy.

---

[12] See, for example, the interesting considerations of P. Coveney, Chaos, entropy and the arrow of time, in: N. Hall, ed., *Chaos – A Science for the Real World,* Italian translation by F. Muzzio, Padua, 203–212, 1992.

Thus entropy breaks the symmetry of time and can change irrespective of changes in energy, energy being a conservative and reversible quantity, whereas entropy is evolutionary and irreversible *per se*. The flow of a non-conservative quantity, negentropy, makes life flow and the occurrence of a negentropy production term is the difference with respect to analysis based on exclusively conservative terms (energy and matter).

The situation is explained in Table 1 "The death of the deer": at the moment of death, mass and energy do not change, whereas entropy does. There is an entropic *watershed* between far-from-equilibrium (living) systems and classical systems (the dead deer or any inorganic non-living system).

Table 1: The death of the deer.

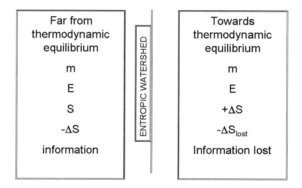

We may conclude that in systems far from thermodynamic equilibrium (biological and ecological), *entropy is not a state function, since it has intrinsic evolutionary properties*, strikingly at variance with classical thermodynamics.

In the framework of evolutionary physics, we must deal with goal functions instead of state functions: ecodynamic models must be based on relations evolving in time; far-from-equilibrium thermodynamics (Prigogine) is the foundation for a new description of nature.

4.5. Self-organizing systems

We observe the bewildering complexity of life in nature. After millennia of human curiosity and centuries of systematic research, we are still far from understanding how this complicated system was established and how it is maintained.

In 1859, Charles Darwin wrote "The Origin of Species", presenting a new evolutionary paradigm. In the same period, Clausius formulated the Second Law of Thermodynamics. Darwin tells us that evolution increases the organization of life. How can this be reconciled with Clausius's theory that disorder accompanies the increasing complexity of nature? This contradiction was magnificently overcome by Prigogine,[13] who observed that non-equilibrium can be a source of order. In fact, irreversible processes may lead to a new type of dynamic state of matter called "dissipative structures". These manifest a coherent, supramolecular character that leads to new, quite spectacular manifestations, such as biochemical cycles involving oscillatory enzymes. These spatio-temporal structures arise from the nonlinear dynamics of such phenomena. How do such coherent structures appear as a result of reactive collisions? Can thermodynamics give us an answer?

The classical thermodynamics of Clausius refers to isolated systems exchanging neither energy nor matter with the outside world. The Second Law then merely ascertains the existence of a function S, which increases monotonically to a maximum at the state of thermodynamic equilibrium:

$$dS/dt \geq 0 \qquad (14)$$

This formulation can be extended to systems that exchange energy and matter with the outside world, in which case it is necessary to distinguish two terms in the entropy change dS: the first, d_eS, is the transfer of entropy across the boundaries of the system; the second d_iS, is the entropy produced within the system by irreversible processes:

$$dS = d_iS + d_eS \qquad d_iS \geq 0. \qquad (15)$$

In this formulation, the distinction between irreversible and reversible processes is essential. Only irreversible processes produce entropy. The Second Principle therefore states that irreversible processes lead to a sort of unidirectional time. For isolated systems $d_eS = 0$ and the previous equations becomes the classical Second Law. Open systems could conceivably evolve to steady states with

$$d_eS = -d_iS \qquad dS = 0 \qquad (16)$$

---

[13] I. Prigogine, *Introduction to Thermodynamics of Irreversible Processes*, C.C. Thomas, Springfield, 1954.

This is a non-equilibrium steady state that should not be confused with thermodynamic equilibrium, and in which order may be created from disorder. Order created in this way no longer violates the laws of thermodynamics; equilibrium is no longer the only attractor of the system, but the world becomes more complex and thermodynamics can embrace new worlds characterized by highly organized as well as chaotic structures.

Prigogine illustrates this entropic behaviour using the concept of entropy production (see Fig. 2).

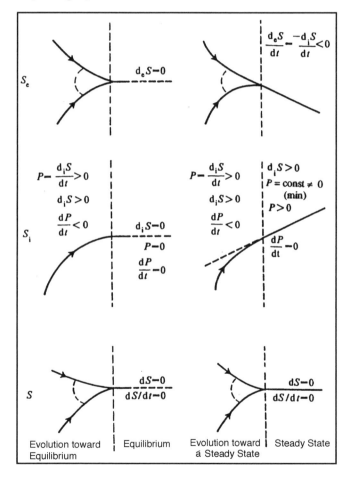

Figure 2: Variations in Entropy in four different cases.

The simplest example of a self-organizing system is the Bénard instability. This argument, together with some new examples of self-organizing systems based on experiments performed in the ecodynamics lab of Siena University, will be discussed in the next Step.

4.6. Entropy and the city

The Second Law of Thermodynamics indicates the paths to avoid in order to maintain life on the Earth. The two paths towards maximum disorder are:

(a) through energy exchanges as heat flows that cancel differences in temperature so that no more work can be done;

(b) through consumption of all resources of an isolated system so that the accompanying increase in internal entropy leads to self-destruction.

Living systems therefore try to avoid thermodynamic equilibrium, keeping themselves as far as possible from that state, self-organizing by exploiting flows of matter and energy from outside and from systems at different temperatures and with different energy conditions.

Economists and our society must heed the Second Law of Thermodynamics. Globalization, "one-way" thought and loss of biological and cultural diversity are leading the world inexorably towards thermal death (or to euthanasia at the hand of entropy, as we call it).

Cities, regions and nations that make a political dogma of isolation, rejecting cultural cross-fertilization and maintaining fundamentalist positions of self-conservation, will meet the same fate. Excessive defence of identity and loss of diversity are two faces of the same thermodynamic foolishness.

Economic systems and cities are thermodynamically sustainable only when they behave as dissipative structures.

As we have seen, the extreme configurations of autarchy and globalization lead to thermal death. Autarchic systems have the same fate as isolated thermodynamic systems; globalization, with its unlimited quantitative growth of flows of energy and matter, levels differences and leads to maximum disorder and maximum entropy.

Socioeconomic systems and cities are living subsystems of the system *biosphere*: they may evolve as do biological systems but their evolutionary processes are faster than environmental responses. This could affect the environment's capacity to support life on Earth indefinitely.

4.7. Georgescu-Roegen

Nicholas Georgescu-Roegen[14] proposed a "fourth law of thermodynamics" for the entropy of matter. Several years ago, we discussed it in Siena with some economist and physicist friends. In the end, everyone stuck to their own opinion.

The physical and mathematical aspects of the dispute were published in an article in *Ecological Economics*[15].

Georgescu-Roegen assumes the existence of entropy of matter that tends to a maximum (maximum disorder and mixing of matter) so that in the end matter is no longer available. According to the proposed fourth law, it is impossible to completely recover matter involved in the production of mechanical work or dispersed by friction, irrespective of the quantity of energy and time expended for recovery. Photosynthesis disproves the hypothesis, because green plants selectively recover carbon dioxide molecules dispersed in the atmosphere, using solar energy. Other examples are the recovery of nitrogen by nitrogen-fixing bacteria in the roots of leguminous plants and the recovery of iron filings by electromagnetic energy. The hypothesis of Georgescu-Roegen is already contained in the Second Law. The point is that to recover dispersed matter, a passage from ordered forms of energy (mechanical, electromagnetic, chemical) to less ordered forms of energy (heat) is required. The rubber worn from tyres or the metal worn from coins can be recovered only at the cost of a great increase in entropy in the surrounding environment (and enormous economic expenditure). In other words, complete recycling of matter is physically possible if a sufficient quantity of energy is available. The problem is that such a waste of energy would lead to a tremendous, certainly unsustainable increase in the entropy of the biosphere.

Essentially, Georgescu-Roegen (not his fourth law) is right. As Maxwell's demon can only exist in an imaginary universe, so the recovery of dispersed particles of matter can only occur in theory, with great expenditure of energy.

---

[14] N. Georgescu-Roegen, *The Entropy Law and the Economic Process*, Harvard University Press, Cambridge, USA, 1971; *Energy and Economic Myths*, Pergamon, New York, 1976.

[15] C. Bianciardi, E. Tiezzi and S. Ulgiati, Complete recycling of matter in the frameworks of physics, biology and ecological economics, *Ecological Economics*, **8**, 1–5, 1993.

5
FIFTH STEP

ORDER OUT OF CHAOS

This revolution is the first genuine creation of improvisation ...
The most perfectly organized chaos in the Universe.

Ernesto Che Guevara[1]

5.1. Science of complexity

The science of complexity, as befits its name, lacks a simple definition. It has been used to refer to the study of systems that operate at the "edge of chaos" (itself a loosely defined concept), to infer structure in the complex properties of systems that are intermediate between perfect order and perfect disorder; or even as a simple restatement of the cliché that behaviour of the systems as a whole can be more than the sum of their parts. Complexity is an inherently interdisciplinary concept that has penetrated a range of intellectual fields from physics to linguistics, but with no underlying, unified theory[2].

Despite all this, the world is indeed made of many highly interconnected parts on many scales, the interactions of which result in complex behaviour that requires separate interpretations of each level. This realization forces us to appreciate the fact that new features emerge as we move from one scale to

---

[1]E. Guevara, Lettera a Ernesto Sábato del 12 aprile 1960, *Ideologie*, **2**,   135–142, La Nuova Italia, Firenze, 1967.

[2] M.M. Waldrop, *Complexity, The Emerging Science at the Edge of Chaos*, Simon and Shuster, New York, 1992.

another, so it follows that the science of complexity is about revealing the principles that govern the way in which these new properties appear[3].

In the past, mankind learned to understand the reality through simplification and analysis on the basis of the idea that a natural system subject to well-defined external conditions will follow a unique course and that a slight change in these conditions will likewise induce a slight change in the system's response. This idea, along with its corollaries of reproducibility and unlimited predictability and hence of ultimate simplicity, has long dominated our thinking and has gradually led to the image of a linear world: a world in which the observed effects are linked to the underlying causes by a set of laws that reduce to proportionality, for all practical purposes.

Nonlinearity gives us new tools to understand complexity and capture the laws behind the exciting variety of new phenomena that become apparent when many units of complex systems interact. Mathematically, the essential difference between linear and nonlinear equations is clear. Any two solutions of a linear equation can be added together to form a new solution: this is the *superposition principle*. The Fourier and Laplace transforms, for example, depend on being able to superpose solutions. Superposition enables systematic methods of solving essentially any linear problem, independent of its complexity. One basically breaks the problem into many small pieces, then adds the separate solutions to get the solution to the whole problem.

In contrast, two solutions of a nonlinear equation cannot be added together to form another solution. Superposition fails. Thus, one must consider a nonlinear problem *in toto*: one cannot break the problem into small sub problems and add their solutions. It is therefore not surprising that no general analytic approach exists for solving typical nonlinear equations. Adding an elementary action to another can induce dramatic new effects in a nonlinear system, reflecting cooperation between the constituent elements. This gives rise to unexpected structures and events, the properties of which can be quite different from those of the underlying elementary laws and may take the form of abrupt transitions, a multiplicity of states, pattern formation or irregular highly unpredictable evolution in space and time, referred to as deterministic chaos. Nonlinear science is the science of evolution and complexity[4].

Physically, the distinction between linear and nonlinear behaviour is best abstracted from examples. For instance, when water flows through a pipe at

---

[3] R. Larter, Understanding Complexity in Biophysical Chemistry, *J. Phys. Chem. B*, **107**, 415–429, 2003.

[4] G. Nicolis, *Introduction to Nonlinear Science*, Cambridge University Press, Cambridge, 1995.

low velocity, its motion is laminar and is characteristic of linear behaviour: regular, predictable and describable in simple analytic terms. However, when the velocity exceeds a critical value, the motion becomes *turbulent*, with localized eddies moving in a complicated, irregular and erratic way that typifies nonlinear behaviour.

In a complex system, we accept that processes that occur simultaneously on different scales or levels are important, and the intricate behaviour of the whole system depends on its units in a non-trivial way. A qualitatively new theory is required to describe the whole system's behaviour, because the laws that describe its behaviour are qualitatively different from those that govern its individual units.

Complex behaviour is ubiquitous in Nature. Let us now look at some examples from physics and biology.

5.2. Thermal convection

Bénard convection is an intensely studied dissipative system, both theoretically and empirically[5,6]. It is therefore a useful starting point for discussing the properties of dissipative structures and for find analogies with development and evolution. The simplifications involved in Bénard cells and the possibilities nonetheless allowed are remarkable. Bénard cells form when a viscous fluid is heated between two plates (to eliminate surface effects) in a gravitational field. Their formation depends on the type of fluid, its depth and the temperature gradient. There is a critical value of the Rayleigh number at which fluctuations in the density of the fluid overcome the viscosity faster than they are dissipated. These fluctuations are amplified and give rise to a macroscopic circular current (Fig. 3).

The Rayleigh number is a dimensionless constant depending on gravitational force, temperature differential from bottom to top, depth of fluid and the coefficients of expansion, viscosity and thermal conductivity of the fluid. Although it is a simple case, the equations governing Bénard convection are highly nonlinear, apparently a prerequisite for the emergence of self-organization[7].

---

[5] S. Chandrasekhar, *Hydrodynamic and Hydromagnetic Stability*, Oxford University Press, Oxford, 1961.

[6] E. Koschmieder, Experimental aspects of hydrodynamic instabilities, in: G. Nicolis, G. Dewel and J.W. Turner, eds., *Order and Fluctuations in Equilibrium and Nonequilibrium Statistical Mechanics*, Wiley, New York, 1981.

[7] E. Bodenschatz, W. Pesch and G. Ahlers, Recent developments in Rayleigh–Bénard convection, *Ann. Rev. Fluid Mech.*, **32**, 709–778, 2000.

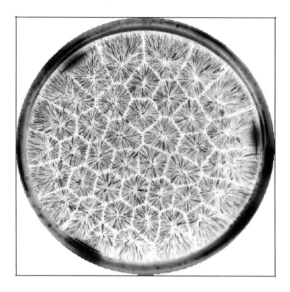

Figure 3: Convection patterns generated by heating a thin layer of viscous fluid in a gravitational field. This phenomenon is known as the Rayleigh–Bénard convection cells.

Bénard convection cells are among the simplest self-organizing phenomena. One aspect of their simplicity is that they depend only on boundary conditions once they reach a steady state. They do not evolve according to an inner logic. Their characteristics are nevertheless instructive. The most salient feature is the emergence of large-scale correlations among molecular movements. Individual molecules move more or less at random, as viewed locally. It is only the statistical average that forms the convection cells. This averaging must have some physical source. In this case it lies in the viscosity of the fluid: no viscosity, no convection. Cohesion within the system is essential, and in Bénard cells this cohesion is caused by viscosity. It also seems necessary that a force be applied throughout the system. In this case, there are two potentials: gravity and temperature gradient. Although heat is applied from outside the system, the temperature gradient exists inside the system in a much stronger sense than the gravitational potential. Unlike the gravitational potential, which is conservative, the temperature gradient induces a non-conservative entropy gradient within the system. Convection is an attempt to reduce this gradient. The temperature differential must be large enough to allow fluctuations sufficient to overcome viscosity faster than the fluctuations are dissipated. This creates an instability in a fluid that is not moving macroscopically.

The necessary conditions for the formation of Bénard cells are: (1) gravitational force and/or pressure gradient (applied force), (2) viscous fluid (internal cohesion), (3) sufficient temperature differential to create fluctuations large enough to overcome viscosity, i.e., internal cohesion of the system on a macroscopic scale (internal entropy gradient).

There is no known way to derive the convecting state from the microscopic movements of the molecules or local fluid dynamics. Bénard cells are genuinely emergent[8].

When a thermal gradient exceeds a critical threshold, the fluid ceases to be at rest and begins to perform bulk movements organized in the form of well-structured convection cells. Adjacent cells rotate in opposite directions. At a given point in space, a small-volume element of fluid may therefore find itself in either of two distinct states, in the sense that it can be part of a cell rotating clockwise or counter clockwise. Before the threshold is reached, the fluid is at rest and beyond the threshold we have two new branches of states.

We observe that nonlinear behaviour associated with multiplicity of states emerges through a bifurcation mechanism (qualitative changes in dynamics are called bifurcations). In addition to this first bifurcation, the system can also undergo a whole series of successive transitions, unveiling practically the entire repertoire of nonlinear behaviours[4].

5.3. Belousov–Zhabotinsky reaction

The Belousov–Zhabotinsky (BZ) reaction belongs to the class of "oscillating chemical reactions", known since the 1920s. The concentrations of the intermediates of these reactions may increase or decrease in time, in a coherent way, producing effects such as sustained oscillations[9]. This behaviour can be translated into a sequence of periodic colour changes, such as blue-red in the BZ system catalysed by ferroin, or transparent-yellow in that catalysed by cerium. These fascinating reactions were a challenge to chemists and scientists in the early 20th century.

---

[8] P. Manneville, *Dissipative Structures and Weak Turbulence*, Academic Press, Boston, 1990.
[9] G. Nicolis and J. Portnow, Chemical oscillations, *Chem. Rev.*, **73**, 365–384, 1973.

The first oscillating chemical reaction was discovered in 1921 by Bray[10], who observed periodic variations in iodide concentrations during decomposition of hydrogen peroxide catalysed by the iodate ion. This behaviour was considered impossible by scientists, since it violated the Second Law of Thermodynamics, and the oscillations in iodide concentrations were attributed to unknown impurities. Bray's observations were not investigated by others and were forgotten.

In 1950, the Russian Chemist B.P. Belousov made a crucial discovery while working on an inorganic analogue of Krebs cycle. He replaced a protein complex catalyst with ceric ions that coloured the solution an intense yellow. To his amazement, the colour of the solution begun to oscillate periodically between yellow and transparent[11].

Once again the scientific community did not accept the experimental evidence because it violated the Second Law of Thermodynamics. Belousov's work was rejected by all scientific journals of the time, and the only trace of his experience can be found in the proceedings of an obscure Russian medical meeting in 1958. Understandably disappointed, Belousov never worked on the reaction again.

Study of the reaction was continued by Zhabotinsky (1964) and is now known as the Belousov–Zhabotinsky reaction. Many who could not accept chemical oscillations cited the Second Law of Thermodynamics. The power of the second law lies in its ability to predict the direction of spontaneous change on the basis of the deceptively simple condition $\Delta S > 0$, where the ΔS is the total change in entropy of an isolated system or of the universe. Applied to chemical reactions, this principle implies that any spontaneous process reaches its final equilibrium state monotonically. Oscillations around a point of equilibrium are not consistent with the Second Law because free energy must decrease monotonically to the equilibrium value. Only since Prigogine has it been possible to understand this complex behaviour[12]. He showed that thermodynamic ideas could be applied to systems far from equilibrium. Prigogine and his co-workers in Brussels focused on chemical systems, showing that a system could organize (decrease its entropy), so long as the net entropy change in the universe was positive. Thus, for example, the

---

[10] W. C. Bray, A periodic reaction in homogeneous solution and its relation to catalysis, *J. Am. Chem. Soc.*, **43**, 1262–1267, 1921.

[11] B.P. Belousov, A periodic reaction and its mechanism, *Sbornik Referatov po Radiatsionni Meditsine*, **145**, 1958.

[12] I. Prigogine, *Thermodynamics of Irreversible Processes*, Wiley, New York, 1955.

concentrations of the intermediates of a reaction can increase and decrease in time while free-energy decreases monotonically as a result of continuous conversion of high free-energy reactants into low free-energy products. Prigogine showed that open systems kept far from equilibrium could spontaneously self-organize by dissipating energy into their surroundings to compensate for the entropy decrease in the system. He called these temporal or spatial structures dissipative structures.

5.4. Mechanism of the Belousov–Zhabotinsky reaction

The BZ reaction consists of the catalytic oxidation of an organic substrate with active methylenic hydrogens (e.g. those of malonic acid) by BrO_3^- in strongly acidic solution. In the last 30 years, many models have been proposed to explain the oscillatory behaviour of this reaction. The first and simplest is known as the Field, Körös and Noyes (FKN model)[13,14,15,16]. The global stoichiometry of the reaction was found to be

$$2BrO_3^- + 3MA + 2H^+ \rightarrow 2BrMA + 4H_2O + 3CO_2 \qquad (17)$$

where MA is malonic acid. The reaction can be catalysed by many redox species (Ce, Fe, Mn, etc.). Here we consider the couple Ce^{III}/Ce^{IV}. The FKN model involves 18 elementary steps and 21 chemicals but can be simplified considering the role of three key species:

- $HBrO_2$ as exchange intermediate;
- Br^- as control intermediate;
- Ce^{4+} as regeneration intermediate.

In this simplified version, the FKN skeleton can be summarised as follows:

---

[13] S.K. Scott, *Chemical Chaos*, Clarendon Press, Oxford, 1991.

[14] I.R. Epstein and A.J. Pojman, *An Introduction to Nonlinear Chemical Dynamics: Oscillations, Waves, Patterns an Chaos*, Oxford University Press, New York, 1998.

[15] R.M. Noyes, R.J. Field and E. Körös, Oscillations in chemical systems. I. Detailed mechanism in a system showing temporal oscillations, *J. Am. Chem. Soc.*, **94**, 1394–1395, 1972.

[16] R.M. Noyes and R.J. Field, Oscillatory chemical reactions, *Ann. Rev. Phys. Chem.*, **25**, 95–119, 1974.

$$Br^- + HOBr + H^+ \rightleftarrows Br_2 + H_2O \tag{R1}$$

$$HBrO_2 + Br^- + H^+ \rightleftarrows 2HOBr \tag{R2}$$

$$BrO_3^- + Br^- + 2H^+ \rightleftarrows HBrO_2 + HOBr \tag{R3}$$

$$2HBrO_2 \rightleftarrows HOBr + BrO_3^- + H^+ \tag{R4}$$

$$BrO_3^- + HBrO_2 + H^+ \rightleftarrows 2BrO_2^\cdot + H_2O \tag{R5}$$

$$Ce(III) + BrO_2^\cdot + H^+ \rightleftarrows Ce(IV) + HBrO_2 \tag{R6}$$

$$Ce(IV) + BrO_2^\cdot + H_2O \rightleftarrows Ce(III) + BrO_3^- + 2H^+ \tag{R7}$$

$$Br_2 + CH_2(CO_2H)_2 \rightleftarrows BrCH(CO_2H)_2 + Br^- + H^+ \tag{R8}$$

$$6Ce(IV) + CH_2(CO_2H)_2 + 2H_2O \rightleftarrows 6Ce(III) + HCO_2H + 2CO_2 + 6H^+ \tag{R9}$$

$$4Ce(IV) + BrCH(CO_2H)_2 + 2H_2O \rightleftarrows 4Ce(III) + HCO_2H + Br^- + 2CO_2 + 5H^+ . \tag{R10}$$

The mechanism can be divided into three main processes that interact with each other.

Process A involves the elementary steps (R1), (R2) and (R3), during which the system consumes Br^- ions according to the global reaction [(R2)+(R3)+3(R1)]:

$$BrO_3^- + 5Br^- + 6H^+ \rightarrow 3Br_2 + 3H_2O. \tag{18}$$

The bromous acid formed in reaction (R3) is rapidly consumed by reaction (R2), keeping the concentration of $HBrO_2$ at a low level. When the concentration of Br^- (consumed in process A) is below a critical value $[Br^-]_{cr}$, process B begins.

Process B involves the elementary steps (R4), (R5) and (R6). The major feature of this process is the oxidation of Ce^{3+} to Ce^{4+} by the radical $BrO_2^\cdot$ which is an intermediate of the autocatalytic production of $HBrO_2$. The global stoichiometry of the autocatalytic process is (R5) + 2(R6):

$$BrO_3^- + HBrO_2 + 2Ce^{3+} + 3H^+ \rightleftarrows 2HBrO_2 + 2Ce^{4+} + H_2O. \tag{19}$$

Exponential increase of $[HBrO_2]$ is limited by its disproportionation according to reaction (R4). The total reaction of process B is given by the sum 2(R5)+4(R6)+(R4):

$$BrO_3^- + 4Ce^{3+} + 5H^+ \rightarrow HOBr + 4Ce^{4+} + 2H_2O \tag{20}$$

At this point, in order to have oscillations, the system must return to process A, the requisite for which is to have a high [Br⁻], in particular $[Br^-] >$ $[Br^-]_{cr}$, and moreover Ce^{4+} must be reduced to Ce^{3+} to be available for process B. These conditions are provided by process C through involvement of an organic substrate (in our case malonic acid).

Process C, according to the original interpretation by Field, Körös and Noyes, HOBr brominates MA, forming Br_2 and bromomalonic acid BrMA. Both MA and BrMA react with Ce^{4+} to yield the reduced form of the catalyst and, in the case of BrMA, bromide ions

$$2Ce^{4+} + MA + BrMA \rightarrow fBr^- + 2Ce^{3+} + \text{other products} \quad (21)$$

where the stoichiometric factor f is the number of bromide ions formed during reduction of the catalyst. The global effect of process C is to produce bromide ions and to restore the reduced form of the catalyst, which resets the cycle and brings the system back to process A. Cyclic repetition of the three processes explains the oscillatory behaviour of the BZ system.

With the FKN model and the appropriate values of the reaction rate constants, it was possible to set up a mathematical model to predict the experimental conditions under which oscillations may be observed.

5.5. Complexity in biology

A typical manifestation of nonlinear behaviour in biology is self-sustained oscillations[17]. Examples can be found at all levels of biological organization, from molecular to supracellular, or even social, with periods of oscillation ranging from seconds to days or years. The beating of the heart is the most obvious of the rhythms that sustain life. Other examples are no less important: B-cells in the pancreas produce insulin in squirts; these periodic pulses are themselves associated with rhythmic oscillations in calcium concentrations. Rhythmic behaviour also arises at metabolic network level: it is a little-known but very interesting fact that hormone levels in the human bloodstream peak five or six times a day, providing an oscillation in key control species for many physiological processes.

Generation of patterns and shapes is a ubiquitous phenomenon in nature, intimately linked with manifestations of oscillatory behaviour. It occurs from microscopic (e.g. self-organization in cells) to macroscopic level (e.g. the mantle of certain animals) in living and non-living systems alike. At every

---

[17] J.D. Murray, *Mathematical Biology*, Springer-Verlag, New York, 2002.

level of complexity, common features seem to relate ecosystem dynamics with the development or organization of biological and chemical systems (the prey-predator model shares identical characteristics with chemical oscillators). Understanding pattern generation is thus inherent to understanding the notion of complexity in open systems[17,18].

Stationary (or slowly varying) patterns, that occur in morphogenesis, and time-dependent patterns, which involve propagating waves, are an important example of structures generated by chemical and biochemical systems. The Belousov–Zhabotinsky reaction seems to have both types of patterns, depending on specific conditions[19,20].

The basic mechanism of morphogenesis, namely differentiation of tissues from stem cells, was proposed 50 years ago by Turing[21], who pointed out that if morphogens obey a reaction-diffusion equation, they may undergo symmetry-breaking transitions. This, in turn, generates spatially organized states, now known as *Turing structures*, which may explain the initial stages of cell aggregation and development. In particular, if we consider a generic reaction-diffusion system

$$\frac{\partial C_i}{\partial t} = D_i \nabla^2 C_i + f_i(..., C_i, ..., C_j, ...) \tag{22}$$

where C_i is the concentration of the ith species participating in the reaction, f_i is the nonlinear kinetics equation describing the reaction and D_i is the diffusion coefficient of species i. Assuming that eqn (22) has a homogeneous steady state solution, $f_i(C_s) = 0$, let us consider the evolution of a small perturbation c_s around C_s and separate it in Fourier space,

$$c_s = \sum_k a_k e^{\lambda_k t + i\vec{k}\cdot\vec{r}} \tag{23}$$

where λ_κ is the growth rate of the mode with a wave vector. Substituting eqn (23) in eqn (22) and retaining only the linear term, we obtain an eigenvalue equation for λ_κ for the linear operator

---

[18] P. Ball, *The Self-Made Tapestry: Pattern Formation in Nature*, Oxford University Press, Oxford, 1999.

[19] J.J. Tyson and P.C. Fife, Target patterns in a realistic model of the Belousov–Zhabotinsky reaction, *J. Chem. Phys.*, **73**, 2224–2237, 1980.

[20] V.K. Vanag and I.R. Epstein, Pattern formation in a tunable medium: the Belousov-Zhabotinsky reaction in an aerosol OT microemulsion, *Phys. Rev. Lett.*, **87**, 228301-1–228301-4, 2001.

[21] A.M. Turing, The chemical basis of morphogenesis, *Phil. Trans. R. Soc. London B*, **327**, 37–52, 1952.

$$L_{ij} = F_{ij} - D_i k^2 \delta_{ij} \tag{24}$$

where F_{ij} is the Jacobian matrix of the kinetic function f_i. For a certain non-zero mode k, Turing instability occurs when the real part of the eigenvalue λ_κ of operator (24) becomes positive, so that the homogeneous steady state becomes unstable and the system undergoes a transition from the homogeneous state to a patterned state. Turing structures are characterized by a constant wavelength and time-independent behaviour. They are generated when the diffusion coefficients of the morphogens (in the simplest case one activator and one inhibitor) are very different.

Target patterns where pulses are emitted periodically from the same leading centre (pacemaker) have been observed in a host of reaction-diffusion systems that generate chemical oscillations. These systems occasionally produce more complex structures, for example outward rotating spirals[22]. The simultaneous presence of propagating waves and stationary patterns was found when the BZ reaction was carried out in the anisotropic environment of a lipid/water binary system. In order to mimic the biological complexity of a cell environment, the Siena Ecodynamics group studied the BZ oscillator in the aqueous compartment of different phospholipid/water lamellar phases[23,24]. All the lipids studied are major components of plasma membranes. Lipids in water are a good model for biological environments, since different aqueous domains in living systems are usually separated by a membrane with a single phospholipid bilayer as basic structural unit. Moreover, many systems have several bilayers stacked together; examples include myelin structure around nerve axons[25], membranes in retinal rods[26] and the extracellular lipid matrix of the stratum corneum[27]. The results of our work were compared with two interesting examples of true biological systems, where similar patterns are

[22] R. Kapral and K. Showalter, eds., *Chemical Waves and Patterns*, Kluwer, Dordrecht (NL), 1995.

[23] A. Magnani, N. Marchettini, S. Ristori, C. Rossi, F. Rossi, M. Rustici, O. Spalla and E. Tiezzi, Chemical waves and pattern formation in the 1,2-dipalmitoyl-sn-glycero-3-phosphocholine/water lamellar system, *J. Am. Chem. Soc.*, **126**, 11406–11407, 2004.

[24] N. Marchettini S. Ristori, F. Rossi and M. Rustici, An experimental model for mimicking biological systems: the Belousov–Zhabotinsky reaction in lipid membranes, *Int. J. Ecodynamics*, **1**, 55–63, 2006.

[25] A.D. Davis, T.M. Weatherby, D.K. Hartline and P.H. Lenz, Myelin-like sheaths in copepod axons, *Nature*, **398**, 571–571, 1999.

[26] E.A. Dratz and P.A. Hargrave, The Structure of rhodopsin and the rod outer segment disk membrane, *Trends Biochem. Sci.*, **8**, 128–131, 1983.

[27] E. Sparr and H. Wennerström, Responding phospholipid membranes – interplay between hydration and permeability, *Biophys J.*, **81**, 1014–1028, 2001.

thought to originate from the propagation of chemical waves. They are illustrated in Fig. 4(b) (mammal cerebral cortex[28]) and 5(b) (bacterial colonies of the slime mould *Dictyostelium discoideum*[29]) and compared with structures obtained in lipid water systems (Figs 4(a) and 5(a), respectively).

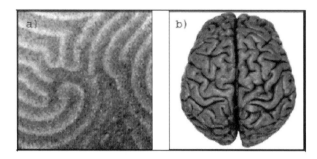

Figure 4: (a) Labyrinthine structures in lipid/BZ system. (b) Human brain. The convoluted ridges (gyri) and valley (sulci) that form the cerebral cortex are thought to originate from a reaction-diffusion process during the fetus development.

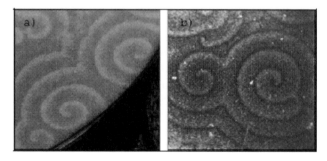

Figure 5: (a) Inwardly rotating spirals in lipid/BZ system. (b) A colony of Dictyostelium Discoideum. The spirals shown are caused by the movement of cells in response to cAMP (cyclic-Adenosine-Mono-Phosphate) waves periodically initiated by the aggregation centre.

---

[28] J.H.E. Cartwright, Labyrinthine Turing pattern formation in the cerebral cortex, *J. Theor. Biol.*, **217**, 97–103, 2002.
[29] P. Foerster, S.C. Müller and B. Hess, Curvature and spiral geometry in aggregation patterns of Dictyostelium discoideum, *Development*, **109**, 11–16, 1990.

Figure 6: Lipid/BZ system with low water content. No interesting patterns or structures have been observed.

Another intriguing result obtained in the Siena lab was the disappearance of all BZ patterns and structures when the water phase/lipid ratio was low. For water concentrations below 70% weight/weight with respect to lipid, no self-organization phenomena were observed (Fig. 6). Since all known forms of life and all physiological solutions have a water content in the range 60–99% (i.e. water is the molecule of life), our findings could open new perspectives in the comprehension of life mechanisms, especially regarding the role of water in self-organizing systems (dissipative structures). This aspect will be treated in more detail in the following paragraph. In conclusion, biorhythms are an important aspect of complexity in biology and nonlinear tools are fundamental for understanding the dynamic behaviour of such systems. Rhythms or oscillations may arise at any level of nature[30]. We usually think of these levels as going from subatomic particles upward: Quarks → Atoms → Molecules → Cells → Organisms → Societies → Ecological Systems → Biosphere. This is the traditional reductionist approach to science. Some scientists hope that by breaking the world into

---

[30] S. Strogatz, *SYNC: The Emerging Science of Spontaneous Order*, Hyperion books, New York, 2003.

smaller and smaller pieces they will eventually achieve an understanding that will enable them to provide a workable description of the whole.

The big question is whether there is a unified theory for the ways in which elements of a system organize themselves to produce behaviours typical of large classes of systems. In complex systems, we accept that processes occurring simultaneously at different scales or levels are important, and the intricate behaviour of the whole system depends on its units in a non-trivial way. Description of the whole system's behaviour requires a qualitatively new theory because the laws governing this behaviour are qualitatively different from those governing the individual units. For example, who could find the law governing the intricate pattern of electrical activity produced by the brain by studying a single neuron? In complex systems randomness and determinism are both relevant to the system's overall behaviour. Complex systems exist on the edge of chaos: their behaviour may be almost regular but may change dramatically and stochastically in time and/or space as a result of small changes in conditions.

Returning to the above scheme, it is evident that universal dynamic behaviours (bistability, oscillations, chaos, waves, etc.) can be observed at many levels, including a top planetary level. These phenomena transcend the traditional organizational hierarchy.

5.6. The supramolecular structure of water

Water is the most abundant substance on Earth and has been extensively studied; a number of model structures have been proposed and refined. However, it remains an anomalous liquid in as far as no single model explains all of its properties.

Much work has been invested in developing models for individual water molecules for use in molecular dynamics simulations.

Mechanically strong random-network models have been proposed, and other models for ice structures.

Dodecahedral water clusters have been reported at hydrophobic and protein surfaces, where low-density water with stronger hydrogen bonds and lower entropy has been found.

The hypothesis advanced here is based on the formation of dissipative structures (supramolecular structures of water structurally similar to a liquid crystal) through hydrogen bonds. The supramolecular structure could be a macromolecule typical of living organisms (plant or animal). We know that these macromolecules are ~80% water, have complex conformations induced initially by an active principle, and self-organize in response to an input of energy (e.g. dynamization or an electromagnetic wave).

A liquid crystal or liquid cluster can be defined as a substance that flows as a liquid but has some order in the arrangement of its molecules; its phase is intermediate between liquid and crystalline solid. Its molecules are typically rod-shaped and about 25 Å in length. The ordering of liquid crystal molecules is a function of temperature. According to Chandrasekhar[31],

> "The term liquid crystal signifies a state of aggregation that is intermediate between the crystalline solid and the amorphous liquid. As a rule, a substance in this state is strongly anisotropic in some of its properties and yet exhibits a certain degree of fluidity, which in some cases may be comparable to that of ordinary liquid".

For the supramolecular structure of water, we suggest a model based on network formation by mechanically strong water clusters: low-density water with stronger hydrogen bonds and lower entropy.

As pointed out by Prigogine, low-entropy dissipative structures have coherent supramolecular character that leads to spectacular new manifestations, such as biochemical cycles involving oscillatory enzymes. These spatiotemporal structures arise from the nonlinear dynamics of such phenomena.

Various experiments[32,33,34] by Louis Rey and at Naples University have provided new information about the physical nature of the liquid state of water and aqueous solutions, opening new avenues in the study of water as a non-equilibrium system. The evolution of aqueous systems vastly exceeds typical "molecular" time-scales. This, together with very low concentration and electromagnetic-field effects, suggests that water and dilute aqueous solutions can be regarded as self-organizing systems.

At macro-level, the behaviour of water is related to biodiversity, the core of biological evolution.

We assert here that the action of drugs and medicines is mediated by the supramolecular structure of water and that the supramolecular structure of

[31] S. Chandrasekhar, *Liquid Crystals*, Cambridge University Press, 1992.

[32] L. Rey, Thermoluminescence of ultra-high dilutions of lithium chloride and sodium chloride, *Physica A*, **323**, 67–74, 2003.

[33] V. Elia and M. Niccoli, Thermodynamics of extremely diluted aqueous solutions, *Ann. N.Y. Acad. Sci.*, **827**, 241–248, 1999.

[34] V. Elia and M. Niccoli, New physico-chemical properties of extremely diluted aqueous solutions, *J. Therm. Anal. Calorim.*, **75**, 815–836, 2004.

water is related to the formation of dissipative structures and to their self-organization.

The role of water is also related to the interactions with solid surfaces and to a different treatment of the concept of probability: this part will be discussed in the Tenth Step.

6
SIXTH STEP

SONGS AND SHAPES OF NATURE

[Roy]
I've seen things you people wouldn't believe,
....attack ships on fire off the shoulder of Orion,
I watched β-beams glitter in the dark near the Tannhäuser gate,
All those moments will be lost in time, like tears in rain...
...time...to die.

Blade Runner, 1982

6.1. The vocal cords, a wonderful dissipative system

All mammals are united by the possibility of emitting sounds and by the mechanism of sound production: the vocal cords. Despite the evolutionary distance separating the various species, the vocal apparatus remains almost unchanged. Sound is produced when the vocal cords are excited by lung pressure controlled by the muscles of the glottis; sound production is controlled by opening and closing the vocal membranes (see Fig. 7).

The vocal folds have been extensively studied in the last 30 years in order to build models for voice synthesizers and sound systems. In 1972, Ishizaka and Flanagan[1] proposed the two-mass model (see Fig. 8).

---

[1] K. Ishizaka and J. Flanagan, Synthesis of voiced sounds from a two mass-model of the vocal chords, *Bell Syst. Tech. J.*, **51**, 1972.

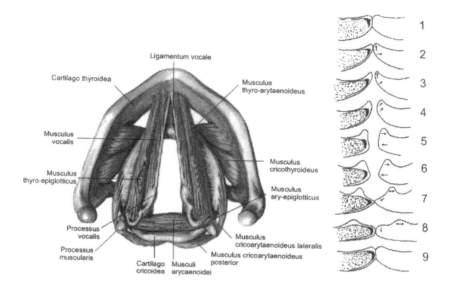

Figure 7: The human vocal cords and their oscillation cycle.

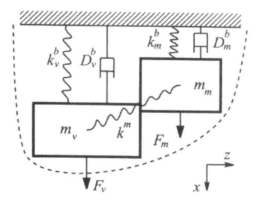

Figure 8: Mechanical model of the vocal cords – a typical viscoelastic dissipative oscillator.

The vocal apparatus is likened to a pair of coupled viscoelastic oscillators excited by lung pressure, moving according to Newton's Second Law as described by the equations:

$$F_{x,v} = m_v \ddot{x}_v + D_v \dot{x}_v + k^m (x_v - x_m)$$
$$F_{x,m} = m_m \ddot{x}_m + D_m \dot{x}_m + k^m (x_m - x_v)$$

(25)

The forces acting on the masses are directly proportional to lung pressure. The pressure on the walls of the vocal tract was computed by Herzel[2]:

$$P = P_s - 1.37 P_s \left(\frac{1}{0.37 - A_1^2 / A_2^2} \right) \Theta(A_1) \Theta(A_2)$$

(26)

where P_s is lung pressure, A_1 and A_2 are the vocal tract areas and $\Theta(\bullet)$ is the Heaviside step function (indicating that there are no active forces when the vocal tract is closed).

Classical analysis of the mechanical model includes linearization of the system and computation of the natural oscillation frequencies

$$\omega_{1,2} = \sqrt{\frac{k_1 + k}{2m_1} + \frac{k_2 + k}{2m_2} \pm \sqrt{\left(\frac{k_1 + k}{2m_1} + \frac{k_2 + k}{2m_2} \right)^2 + \frac{k^2}{m_1 m_2}}}.$$

(27)

From the point of view of classical science, it is possible to reconstruct or derive vocal cord motion by solving the system of differential equations (25) and appropriately modelling the parameters; it is therefore theoretically possible to reproduce any mammalian voice. Fortunately, nature is not so flat and the nonlinear terms introduced in the equations of motion (together with those in the real system that cannot be modelled) make the dynamics of the system completely unpredictable and non-reproducible. The mathematical equations do not contain the wonderful diversity of sounds produced in nature.

What we can do with our instruments is nothing compared to nature. The theories of nonlinear dynamic systems may nevertheless help us to understand (but not to simulate) the behaviour of nature. Using *nonlinear*

---

[2] H. Herzel, Bifurcations and chaos in voice signals, *Appl. Mech. Rev.*, **46**, July, 1993.

time series analysis methods, we can peer into the folds of nonlinearities, in search of something that divides and joins at the same time.

6.2. Looking for the "voice fingerprint"

The characteristic parameters of the model can be divided into two groups, those, such as air density, which do not depend on the individual and those, such as muscle tension, the shape of the glottal apparatus, the dimensions of the vocal cords, and so on, that depend on the speaker.

The experimental results of the last 30 years have revealed many observed behaviours, ranging from regular oscillations through aperiodic motion to fully chaotic behaviour. In particular, the chaotic phenomena observed in human and animal vocalizations may be attributed to a hidden source of information situated between determinism and noise.

The vocal cords behave like a dissipative oscillator and it is possible to distinguish two simultaneous motions: a regular oscillation, characterized by periodic behaviour that determines the fundamental frequency of the vocalization, and an aperiodic motion overlapping the main oscillations. The first type of motion does not depend on the speaker and is the result of mechanical oscillations of the vocal cords; this motion can be simulated by integrating the equations of motion. The sound produced is non-human and inexpressive. The failure of many synthesizers from the point of view of this theory is simple: aperiodic motion is not noise, but it is a distinguishing character of an individual, it is the "*vocal fingerprint*".

The problem is to identify the fingerprint. Vocalization, which comes from living individuals, is an irregular phenomenon and it is not clear or well studied how to extract this subtle information. The analysis can be based on a single observation of the complex dynamics giving rise to the voice: the time series derived from a recording of the vocalization. According to the theories of nonlinear time series analysis, it is possible to reconstruct the overall behaviour of the vocal apparatus and learn something about voice production mechanisms.

6.3. Vocal fingerprint as biodiversity indicator

The attractor of the underlying dynamics can be reconstructed by the Takens theorem (see Appendix A) starting from a single observed variable. If the voice of an individual is different, the dynamics of its vocal folds are also different and the shape of the attractor must be different. Application of the method of delays to bat vocalizations by the Ecodynamics research group of Siena University has produced extremely interesting results.

The vocal fingerprint is a unique characteristic of any individual, though all mammals (not only humans) have basically the same vocal apparatus. If extracted, the vocal fingerprint is a powerful biodiversity indicator since no two voices, and therefore their dynamics, are identical. Voice production is not only a mechanical phenomenon but also depends on individual internal factors, not well known or reproducible.

From the point of view of biodiversity defence, this is an important step: biodiversity exists and can be measured, it is no longer an abstract concept but a measurable thing.

6.4. Bats

Bats are mammals. The two-mass model is valid for their vocal apparatus and all the above techniques can be applied. The vocalization emitted by bats is a "pitch" used as echo-sonar. It is very short and intense (see Fig. 9).

We can now make the following statements:

(1) The frequency of the pitch is very high (50 KHz). These frequencies can only be produced by small animals. Equation (27) shows that the mass of the vocal cords is inversely proportional to the frequency.

(2) Bats are classified on the basis on their size and on the shape of the echo-locating pitch. Certain software makes a non-accurate classification of species based on the shape of the pitch spectrogram.

Time-series analysis techniques can be used to extract further information, and the three main steps are as follows.

(1) Computation of the *embedding dimension*, in order to find the dimensionality of the system.

(2) *Attractor reconstruction* using the method of delays expressed in the Takens theorem. The shape of the attractor, said to be "strange" if chaotic, is very important; it reflects system dynamics in a time-independent space, and reveals certain invisible dynamic patterns. In particular, the reconstructed attractor of a vocal signal is directly related to the motion of the vocal folds.

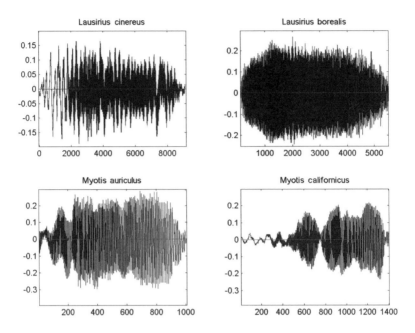

Figure 9: Echolocation pitches of different genera and species of bat.

(3) *Computation of the Lyapunov exponents and the maximum Lyapunov exponent (MLE).* The MLE tells us how sensitive the dynamic system is to initial conditions and discriminates system "chaoticity".

The dynamics of the vocal folds vary from species to species, and in a more subtle way from one individual to another. Visual inspection of the attractors of pitches emitted by bats of different genera reveals detectable differences. Different genera have different attractor topology: the shape of the attractor is the first evidence of the vocal fingerprint (see Figs 10 and 11).

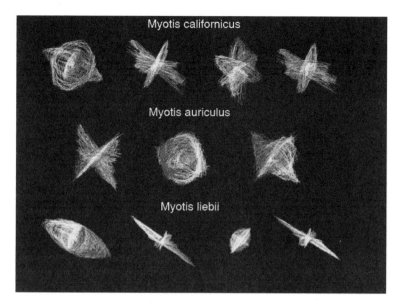

Figure 10: Reconstructed attractors of the genus *Myotis*.

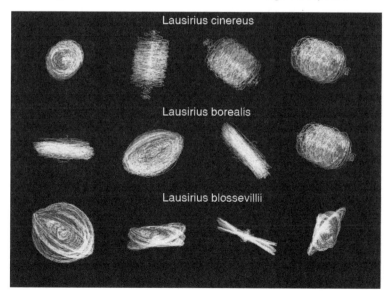

Figure 11: Reconstructed attractors of the genus *Lausirius*.

6.5. Biodiversity monitoring: the Monte Arcosu deer

The last remaining populations of a sub-species of the red deer, the Sardinian deer Cervus elaphus corsicanus, are found in the well-conserved evergreen forest of Monte Arcosu in Sardinia. Cervus elaphus is the largest and most phylogenetically advanced species of Cervus. Head and body length is 1.65–2.65 m, tail length 0.11–0.27 m, height at the shoulder 0.75–0.15 m and weight 75–340 kg. The largest and strongest male generally has the largest harem. In order to maintain this position of superiority he must keep rival males away from his females by bellowing and intimidation. After vocalizing, the largest males size each other up, and if antler and body size are similar, they lock their antlers and battle it out in a pushing match.

This species, once widespread throughout the Mediterranean, is now threatened. With the help of WWF Italy, the Ecodynamics group in Siena obtained some recordings of different males in the oasis population. The vocalizations were analyzed and the results are shown in Fig. 12, together with a typical strange attractor (see Fig. 12(b))[3].

The spectrogram in Fig. 13 provides further proof of the validity of the two-mass model. Although the signal is very irregular, a fundamental frequency of 70 Hz can be identified (see Fig. 13(c)). This low frequency is explained by equation (27) since the vocal cords of the deer are large. The spectrogram also shows that the fundamental frequency is immersed in something similar to noise, but visual inspection of the attractors shows a fully structured pattern typical of chaotic dynamics. Since this is the voice of an animal, it seems unlikely that the irregular pattern is noise[4]. From the point of view of attractor shape, we cannot say anything more about the vocal fingerprint, but the chaotic nature of the motion can be measured by the maximum Lyapunov exponent. When this was performed for different individuals, each animal was found to have a specific MLE; this is a preliminary but very interesting result.

Besides the shape of the strange attractor, we can add another parameter to the vocal fingerprint: the MLE.

---

[3] A. Facchini, S. Bastianoni, N. Marchettini and M. Rustici, Characterization of chaotic dynamics in the vocalization of Cervus elaphus corsicanus, *J. Acoust. Soc. Am.*, **114**, 3040–3043, 2003.

[4] I. Tokuda, T. Reide, J. Neubauen, N.J.Owren and H. Herzel. Nonlinear analysis of irregular animal vocalizations, *J. Acoust. Soc. Am.*, **111**, 2908-2919, 2002.

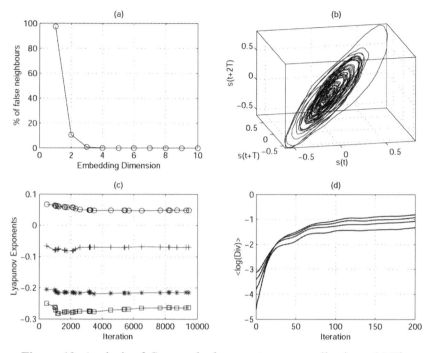

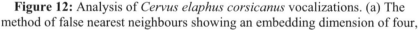

Figure 12: Analysis of *Cervus elaphus corsicanus* vocalizations. (a) The method of false nearest neighbours showing an embedding dimension of four, a result in line with the dimension of the two-mass model system. (b) Reconstructed attractor, a very good example of strange attractor. (c) Computation of all the Lyapunov exponents by means of the Sano and Sawada algorithm. (d) Computation of the maximum Lyapunov exponent by the Rosenstein method; the value MLE = 0.48 indicates that the vocalization is chaotic.

6.6. Applications in medicine: analysis of the crying of newborns

Another important field of application is voice analysis in medicine. The Ecodynamics research group and the Neonatology unit of Siena University are studying the cries of newborns with health problems in order to evaluate the pain suffered by the newborns by means of the time series analysis techniques.

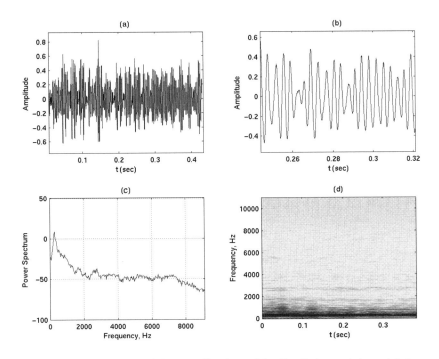

Figure 13: Time series of the vocalization of the Sardinian red deer: (a) the complete series; (b) the time series analyzed; (c) spectrum of the series, with a peak at 70 Hz, the fundamental frequency; (d) spectrogram of the series showing irregular components of the signal: the fundamental frequency seems immersed in noise.

This type of information is very important in medicine: crying is an early form of communication for newborns and the literature distinguishes different forms of crying associated with hunger, anger, pain and so forth. The pain suffered by newborns is fundamental in the work of neonatal specialists. Recent studies[5,6] have endeavoured to distinguish different types

---

[5] C.V. Bellieni, R. Sisto, D.M. Cordelli and G. Buonocuore, Cry features reflect pain intensity in term newborns: an alarm threshold, *Pediatric Res.*, **55**, 142, 2004.

[6] W. Mende, H. Herzel and K. Wermke, Bifurcations and chaos in newborn infant cries, *Phys. Lett. A*, **145**, 1990.

of crying and a scale known as DAN (Doleur Aiguë du Nouveau Né), in which various principal aspects of crying are scored, has been developed. However, the role of crying has rarely been considered.

Using the nonlinear time-series analysis approach, the following points can be considered regarding the two-mass model:

(1)   since the vocal cords and the vocal apparatus of newborns are small, a high sound frequency can be expected;

(2)   since the vocal fold tissues are young and very elastic, over-excitation may produce nonlinear oscillations;

(3)   since pain modifies psycho-physical behaviour, a painful cry may be different from another.

Visual inspection of the time series and spectrograms of Figs 14 and 15 confirm this hypothesis[7]:

- when the DAN score is low, the time series of the cry is periodic;

- with increasing pain, the time series is characterized by regular and irregular oscillations, as with a toroidal transition in dissipative systems;

- when the DAN score is very high, the time series is completely irregular and the spectrogram shows patterns typical of chaotic motion.

This behaviour is typical of dissipative systems into a cascade bifurcations: pain is therefore a "bifurcation parameter".

6.7. Putting time into attractors: the recurrence plots

Reconstruction of the attractor is a good technique to inspect the dynamics of a system, but in real time series, when the signal is non-stationary and the embedding dimension is greater than three, visual inspection would require a multidimensional graph, but we are limited in space!!

Furthermore, the attractor does not give information about the temporal evolution of the curve in phase space, which would be an advantage for biological signals.

---

[7] A. Facchini, C.V. Bellieni, N. Marchettini, F.M. Pulselli and E.B.P. Tiezzi, Relating pain intensity of newborns to onset of nonlinear phenomena in cry recordings, *Phys. Lett. A*, **338**, 332–337, 2005.

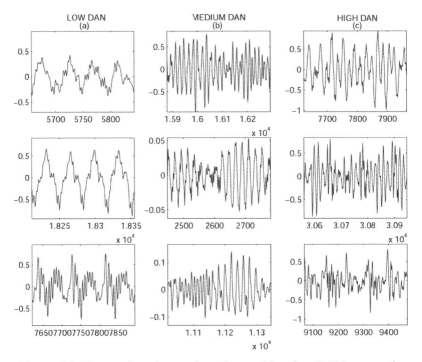

Figure 14: Time series of several newborns: (a) at low DAN scores, the signal is complicated but periodic; (b) at medium DAN scores, the signal is intermittent regular–irregular; (c) at high DAN scores, the signal is completely irregular, typical of chaotic motion.

In 1987 Eckmann[8] proposed a tool for visual investigation of high-dimensional dynamics through visualization in a two-dimensional binary plot. The basic assumptions were that if the dynamics belongs to a deterministic system, the curve in phase-space will pass through the same regions at successive times. This behaviour is known as *recurrent.* To build a recurrence plot (RP), we consider every point on the attractor curve and the distance δ between each point and all others; if δ is below a threshold ε, then the two points i and j are said to be *recurrent,* and a point is placed at (i, j) in the binary diagram, mathematically expressed as

[8] J.P. Eckmann, S.O. Kamphorst and D. Ruelle, Recurrence plots of dynamical systems, *Europhys. Lett.*, **4**, 973, 1987.

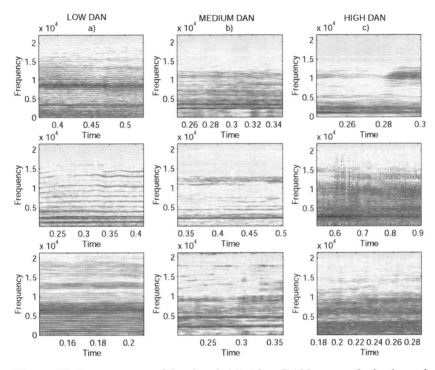

Figure 15: Spectrograms of the signal. (a) at low DAN scores, the horizontal lines represent the harmonic frequencies of the signal, the signal is periodic because the horizontal lines are present for the whole time; (b) at medium DAN scores, some harmonics disappear, (c) at high DAN scores, the motion is completely irregular and only the fundamental frequency can be detected.

$$R_{i,j}^{m,\varepsilon} = \Theta\left(\varepsilon - \left\| x_i - x_j \right\|\right) \tag{28}$$

where $x_{i,j} \in \Re^m$ and ε is the threshold. Since any state is recurrent with itself, the RP has a $\pi/4$ diagonal, also known as the line of identity (LOI). To compute an RP, a norm must be defined, and the most widely used are L_1, L_2, L_∞. The latter is the most popular because it is independent of phase space dimensions and no rescaling is required to compare different RPs. Furthermore, particular attention is needed in the choice of ε. To estimate it, noise levels must be considered and suggested values are some percentage of the maximum diameter of the attractor (but not greater than 10%).

Recurrence plots are thus characterized by different textures and patterns, indicating the nature of the dynamics; these patterns are known as *typology* and *textures*.

Typology provides an overall impression and may be:

- *Homogeneous*, typical of stationary processes, and usually associated with white noise;
- *Periodic*, diagonal lines parallel to the LOI are typical of periodic systems;
- *Drift* behaviours are caused by slowly varying parameters in the system, and changes in dynamics show as white areas or bands.

Textures are composed of local structures that can be observed in a RP, namely, single points, diagonal lines, vertical and horizontal lines

- *Single points* the state does not persist for a long time; usually a RP made only of single points is related to white noise.
- *Diagonal lines* this kind of texture can be expressed by

$$R_{i+k,j+k} = 1\big|_{k=1}^{L}$$

  where L is the length of the line. This indicates that the trajectory passes through the same region of phase space at different times.
- *Vertical and horizontal lines* of length L can be expressed by

$$R_{i,j+k} = 1\big|_{k=1}^{L}$$

  indicating that the state of the system does not change or changes slowly in time.

Fig. 16 shows typical patterns observed in RPs.

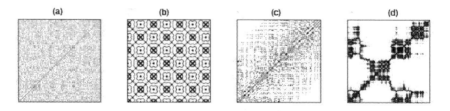

Figure 16: Examples of patterns observed in RPs of different types of signal: (a) white noise, (b) periodic signal; (c) drifting signal; (d) chaotic non-stationary signal.

The methodology we have just described provides a new way to explore complex dynamics, since it enables hidden temporal recurrences in the attractor to be investigated. This is especially true of voice signals, which are full of irregularities due to the desynchronized motion of the vocal folds. The complex nature of the vocal apparatus means that oscillations show temporal recurrences that differ from one individual to another and that can be observed by visual inspection of the RP.

6.8. Investigation of biodiversity in gibbon vocalizations

We now consider calls vocalized by gibbons *Nomascus concolor*, *Nomascus gabriellae*, *Bunopithecus hoolock*, *Hylobates klossii* and *Hylobates lar*.

The calls, technically called "notes" are part of a larger vocalization structure known as a "song", as defined by Thorpe[9]: "*What is usually understood by the term song is a series of notes, generally more than one type, uttered in succession and so related as to form a recognizable sequence of pattern in time*". In a gibbon song, a note is any single continuous sound of any distinct frequency or frequency modulation produced during inhalation or exhalation.

Calls were recorded at a sampling frequency of 11025 Hz and preliminary analysis with spectrograms (1024 Fast Fourier Transform points) did not show patterns typical of irregular signals (see Fig. 17).

Phase-space dynamics were reconstructed considering an embedding dimension $D_E = 8$ and a delay time $T_D = 3$ for all signals, while RPs were computed considering the maximum norm with a fixed neighbourhood size threshold. The embedding dimension was computed considering the dimensionality of the two-mass model ($D_E = 4$) doubled by overembedding[10], while the delay time was chosen at the first minimum of the average mutual information function.

The results of the spectrogram-based analysis (see Fig. 17) revealed that the signals were almost periodic over the entire window, and were characterized by slight amplitude modulation. Only in the last part of the *Hylobates lar* call could a slight frequency modulation be seen.

---

[9] W.H. Thorpe, *Bird-Song. The Biology of Vocal Communications and Expression in Birds*, vol. 12 in: *Monographs in Experimental Biology*, Cambridge University Press, Cambridge, UK, 1961.

[10] R. Hegger, H. Kantz, L. Matassini and T. Schreiber, Coping with nonstationarity by overembedding, *Phys. Rev. Lett.*, **84**, 4092–4095, 2000.

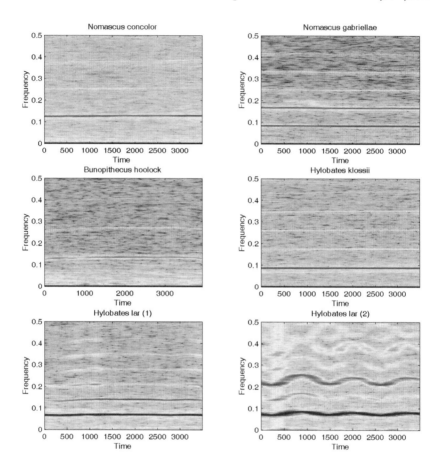

Figure 17: Spectrograms of gibbon calls. The signals seem completely periodic; only the last call of *Hylobates lar* (2) shows a slight frequency modulation.

Visual inspection of the RPs revealed a surprising characterization of the calls. The patterns in Fig. 18 differed widely from species to species.

The two *Hylobates lar* calls are most interesting. Both are characterized by slight frequency modulation and the spectrograms do not provide any further information. On the other hand, the RPs look completely different, since we are now exploring the spatio-temporal recurrences inside the underlying dynamics of the motion of the vocal folds. Is this the vocal fingerprint?

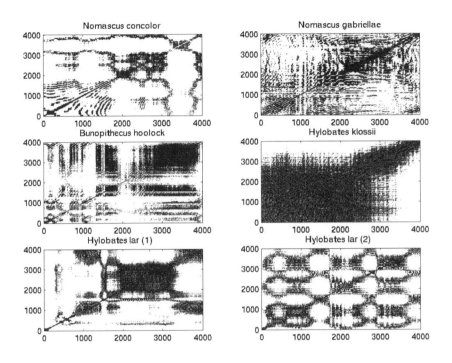

Figure 18: Recurrence plots of gibbon vocalizations. Each call is characterized by different patterns and the two *Hylobates lar* gibbons have completely different patterns. Have we found the vocal fingerprint?

The curved lines forming the patterns may be related with hidden information in the signal[11]. A vocal fingerprint can be compared to an actual fingerprint (see Fig. 19).

---

[11] A. Facchini, H. Kantz and E. Tiezzi, Recurrence plot analysis of nonstationary data: the understanding of curved patterns, *Phys. Rev. E*, **72**, 21915, 2005.

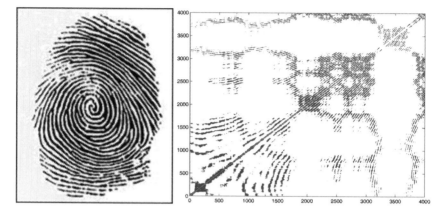

Figure 19: Fingerprint and vocal fingerprint.

6.9. Recurrence plots in medicine: analysis of newborn cries

Since RPs distinguish different types of dynamics better than traditional techniques, application to newborn crying may provide further useful information. RPs of the cries considered in the previous section were analyzed and the results are shown in Fig. 20.

Column (a) shows low-DAN cries and the patterns are typical of periodic signals. When pain increases to medium values, column (b), some white bands appear in the RPs and the patterns are more irregular but still characterized by long lines parallel to the LOI. Column (c) shows high-pain cries and the RPs clearly show patterns typical of chaotic time series with distributions of short parallel lines.

The advantage of RPs is that visual inspection provides much information about evolution of the dynamics and about the frequencies contained in the signal. Since they perform highly with non-stationary signals, they are preferable to spectra or spectrograms, which are better with stationary signals, in these special cases.

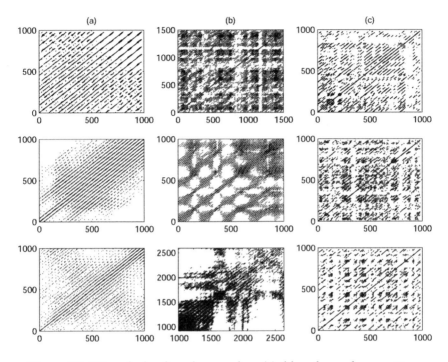

Figure 20: RP analysis of newborn crying: (a) this column shows patterns typical of periodic signals for low DAN scores.; (b) when the DAN score increases, some intermittent patterns can be seen in the RPs and the white bands reveal changes in dynamics not detectable with spectrograms; (c) when DAN scores are high, the RPs are typical of chaotic systems.

Appendix A: Phase space reconstruction

The attractor of the underlying dynamics can be reconstructed in phase space by applying the time delay vector method[12,13]. Starting from a time series $s(t)$ = $[s_1, ...s_N]$, system dynamics can be reconstructed using the delay theorem of Takens and Mane. The reconstructed trajectory X can be expressed as a matrix where each row is a phase space vector

---

[12] E. Ott, *Chaos in Dynamical Systems*, Cambridge University Press, 1993.
[13] H.D.I. Abarbanel, *Analysis of Observed Chaotic Data*, Springer, 1996.

$$\overline{\overline{X}} = [X_1, X_2, ..., X_M]$$

where $X_i = [s_i, s_{i+T}, ...s_{i-(D_E-1)T}]$ and $M = N - (D_E - 1)T$. The matrix is characterized by two key parameters: *embedding dimension D_E* and *delay time T*. The embedding dimension is the minimum dimension for which the reconstructed attractor can be considered completely unfolded and there is no overlapping of reconstructed trajectories. If the chosen dimension is lower than D_E the attractor is not completely unfolded and the underlying dynamics cannot be investigated.

A higher dimension should not be used due to the increase in computational effort. The algorithm used for the computation of D_E is the method of *false nearest neighbours*[14]. A false neighbour is a point of trajectory intersection in a poorly reconstructed attractor. As the dimension increases, the attractor is unfolded with greater fidelity, and the number of false neighbours decreases to zero. The first dimension with no overlapping points is D_E. The delay time T represents a measure of the correlation existing between two consecutive components of D_E-dimensional vectors used in the trajectory reconstruction. Following a commonly applied methodology, the time delay T is chosen at the first minimum of the average mutual information function[15].

Appendix B: Lyapunov exponents

Chaotic systems are sensitive to initial conditions. This profoundly affects the time evolution of trajectories starting from infinitesimally close initial conditions and Lyapunov exponents are a measure of this dependence. These exponents provide a coordinate-independent measure of the local stability properties of a trajectory. If the trajectory evolves in N-dimensional state-space, there are N exponents arranged in decreasing order, known as the *Spectrum of Lyapunov Exponents (SLE)*:

$$\lambda_1 \geq \lambda_2 \geq \cdots \geq \lambda_n.$$

Conceptually, these exponents are a generalization of eigenvalues used to characterize different types of equilibrium points.

---

[14] H.D.I. Abarbanel and M.B. Kennel, Local false nearest neighbours and dynamical dimensions from observed chaotic data, *Phys. Rev. E*, **47**, 3057, 1997.

[15] A.M. Fraser and H.L. Swinney, Independent coordinates for strange attractors from mutual information, *Phys. Rev. A*, **33**, 1134, 1986.

A trajectory is chaotic if there is at least one positive exponent, namely the MLE, the value of which provides a measure of the rate of divergence of infinitesimally close trajectories and of the unpredictability of the system, providing a good characterization of the underlying dynamics.

Starting from the reconstructed attractor X, it is possible to compute the SLE, consisting of exactly $n = D_E$ exponents, by the method of Sano and Sawada[16]. This method is qualitative, and when the exponent, λ_1, is positive, a more accurate method is necessary for the computation.

The method of Kantz[17] is used to compute the MLE from the time series. This method measures, in the reconstructed attractor, the average divergence of two close trajectories at the time $d_j(i)$. This can be expressed as:

$$d_j(i) = C_j e^{\lambda_1(i\Delta t)}$$

where C_j is the initial separation. Taking the logarithm of both sides we obtain

$$\ln d_j(i) = \ln C_j + \lambda_1(i\Delta t).$$

This is a set of approximately parallel lines (for $j = 1,2,...,m$) each with a slope roughly proportional to λ_1. The MLE is easily calculated using least-squares fit to the average line defined by

$$y(i) = \frac{1}{\Delta t} < \ln d_j(i) >$$

where $\langle \bullet \rangle$ denotes the average overall value of j. Fig. 12(d) shows a typical plot of $\langle \ln d_j(i) \rangle$: after a short transition there is a linear region that is used to extract the MLE.

---

[16] M. Sano and Y. Sawada, Measurement of the Lyapunov spectrum from a chaotic time series, *Phys. Rev. Lett.*, **55**, 1082, 1985.

[17] H. Kantz, A robust method to estimate the maximal Lyapunov exponent of a time series, *Phys. Lett. A*, **185**, 77, 1994.

7
SEVENTH STEP

URBAN DYNAMICS: SUGGESTIONS FROM EVOLUTIONARY PHYSICS

Irene is a name for a city in the distance, and if you approach, it changes; there is the city where you arrive for the first time; and there is another city which you leave never to return. Each deserves a different name; perhaps I have already spoken of Irene under other names; perhaps I have spoken only of Irene.

Italo Calvino[1]

7.1. Dissipative structures and cities

Observing a chessboard and the two formations of pawns casually laid out on the board, we study our next move. We try to imagine possible arrangements of the pieces based on a progression of movements composed of lines that may be orthogonal or diagonal, L-shaped or limited in length. In the mind of an expert player, possible moves are devised by weaving potentially limited lines within the limits of the chessboard and the rules of the game. A network of potential relationships must be visualized in the spaces between the pieces in order to elaborate coherent strategies towards the objective of checkmate, the end of the game.

Cities are like a complicated new chessboard, where we find various pawns having multiple relationships and degrees of freedom, orientated towards a series of shared objectives, essentially sustainability. However, sustainability means continuation of the game in time without allowing it to

---

[1] I. Calvino, *Invisible Cities*, Harcourt Brace, New York, 1974.

finish. Like the chess player, the town planner studies possible configurations, selecting the most suitable to satisfy the need for specific or general services. Like the pieces on a chessboard, a city's configuration may appear to be static and the individual parts isolated and disconnected. However, any investigation or planning of a city must consider the relationships between the parts and their connections with the living world. This plot of relationships, like potential moves of pawns, is not manifest and must be imagined in the fabric of the city. The dynamics of its networks (flows of energy, matter, people, goods, information and resources) are fundamental for an understanding of the evolving nature of today's cities.

These necessary conditions are not, however, sufficient for a complete understanding of city systems. A chessboard is a linear system governed by rigid geometric rules capable of generating situations that are complicated but always determinable. An urban system, on the other hand, is both complex and complicated, and therefore not completely determinable. Dynamic phenomena and surprising new events spring from the variable relationships between the parts and from collective behaviours that escape individual control. The evolution of urban systems is governed by irreversible and stochastic processes combining choice and chance. Clear signs may be recognized, for example the way in which certain technologies have recently changed the way people move and exchange information, and how they use public and private urban space. Mobility, telecommunications, energy and material flows that feed a city raise practical evolving questions in the field of urban studies.

Emergent landscapes have changed the perception of cities, drawing attention to general behaviours and ordering principles in a holistic framework. The idea that a city can be conceived as a complex self-adapting system, or even a living ecosystem, is a key point for urban research and opens promising perspectives to direct strategic choices in future development. Certain concepts, from complex system theory to evolutionary physics, from thermodynamics to ecology, confirm this hypothesis and are in line with Odum's[2] definition of an ecosystem as a unit of biological organization consisting of all the organisms in a given area interacting with the physical environment.

The interactions between society and the built environment, living and non-living systems, change and take various forms that can be observed on different space-time scales. Organizational and developmental properties of

---

[2] E. Odum, The strategy of ecosystem development, *Science*, **164**, 262–270, 1969.

cities and ecosystems are considered as wholes and their structural and dynamic properties studied in order to describe and explain the formation of macro-level patterns in systems composed of many interacting micro-level components[3]. An ecosystem approach is required if we assume that some of the properties and behaviours of certain urban systems depend on interactions between their parts and with the surrounding environment. A proper understanding of these properties and behaviours will require bringing these system-environment relations explicitly within the field of investigation[4].

Evolutionary physics and non-equilibrium thermodynamics are the main disciplines for studying dynamics of complex systems and self-organization in urban social systems. Prigogine[5] underlines that cities and living organisms embody different types of functional order:

"To obtain a thermodynamic theory for this type of structure we have to show that non equilibrium may be a source of order. Irreversible processes may lead to a new type of dynamic states of matter which I have called "dissipative structures". [...] These structures are today of special interest in chemistry and biology. They manifest a coherent supermolecular character which leads to new quite spectacular manifestations."

The idea of dissipative structure incorporates the general characteristics and necessary conditions for the development and survival of living non-equilibrium systems. In fact, these structures are defined as living systems that exchange energy and matter with the external environment and self-organize. The ability to self-organize into complex structures and to self-maintain in space and time is aimed at conserving a steady state (dynamicity, diversity, life, etc.), far from thermodynamic equilibrium and with low entropy by virtue of interactions among constituents and systems. How can a city be conceived as a dissipative structure?

Absorption of external input (negentropy) and emission of internal output (entropy, heat, etc.) is a principle that works for social systems, economics, human settlements and all their dynamics. Cities absorb flows of high-quality

---

[3] R.E. Ulanowicz, *Growth and Development: Ecosystems Phenomenology*, Springer-Verlag, New York, 1986.

[4] K. De Laplante and J. Odenbaugh, What isn't wrong with ecosystem ecology, in: R.A. Skipper, C. Allen, R. Ankeny, C.F. Craver, L. Darden, G.M. Mikkelson, and R.C. Richardson, eds., *Philosophy of the Life Sciences*, MIT press, Cambridge, USA, 2006.

[5] I. Prigogine, Time, structures and fluctuations, *Nobel Lecture*, 1977.

energy from the external environment and emit heat, wastes and pollutants; their internal entropy decreases by self-organization in the form of structures, information, social patterns and economy. Resource flows feed these *dissipative cities*, as if they were ecosystems composed of organisms and are metabolized and continuously used to sustain its ordered structure during the time. Cities are physical systems in contact with various sources and sinks; matter and energy flow through them from the sources to the sinks.

Let us consider the flow diagram:

energy sources → *intermediate system (city)* → *sinks*

and divide the system into two parts:

(1)   source(s) + sink(s);
(2)   intermediate system(i).

According to the Second Law of Thermodynamics

$$dS_s + dS_i \geq 0 \qquad (29)$$

where S is the entropy of source sink and S_i is the entropy of the intermediate system. The flow of energy from the source to the sink always involves an increase in entropy

$$dS_s > 0 \qquad (30)$$

whereas the only restriction of the second law on dS_i is that:

$$-dS_i \leq dS_s \qquad (31)$$

so that the entropy of the intermediate system (in our case a city) may decrease if there is an energy flow. A flow of energy provides the intermediate system with quantities of energy for the creation of far-from-equilibrium states, that is, states far from thermal death. The further the non-equilibrium system is from equilibrium, the more ordered it is. If left to itself, the ordered state of a biological system decays to the most disorderly state possible. This is why work must continuously be done to order the system.

For example, the surface of the Earth (intermediate system) receives a flow of energy from the Sun (source) at 5800 K (temperature of the surface of the Sun; the core of the sun is millions of degrees hotter) and returns it to the sink of outer space at 3 K. In this vast temperature range lies the secret of life and the possibility of work against entropic equilibrium, moving living systems away from equilibrium, towards ordered, negentropic, living states. Living systems are maintained in a "steady state", as far as possible from equilibrium, by the flow of energy.

A city is part of a much vaster territory and many processes have a wide range of action or, even, a global dimension. Without continuous flows of input energy that build order, these systems would degrade. The inventory of the inputs that supply a city comes from an observation of multiple dynamics and processes. Their visualization in a diagram gives a synthetic description of resource flows and transformation processes that take place in the territory; it works as a scheme in which relationships between the system and the outside and between its own parts have been decoded in the form of flows of energy and materials. Specifically, an *energetic diagram* provides a comprehensive glance on urban and regional dynamics based on the ability to group different aspects of a territory and different sectors into a unique vision. This *energetic diagram* (Fig. 21) is built according to the graphic directives by H. Odum[6] based on the *energy system symbols*.

According to the Odum's *energy system language*, in an energy diagram the large rectangle defines the boundaries of the system under study, an urban region. In reference to these boundaries, the multiple relations among parts inside and outside the system are classified. Macrosectors (right-smoothed rectangles and small rectangles) are sub-systems with their own specific processes: agriculture, woods and forests (*forests*), cattle-breeding, hunting and fishing (*breeding*) are primary transformation processes; thermal production of electricity (*thermoelectricity*), manufacturing industry, craft and tertiary are secondary transformation processes. Sub-systems, such as *city and population,* are described as consumers in the system (hexagons) that work as attractors for those fluxes that supply urban areas, where in general consumptions are mostly localized. Inputs to the system come directly from the environment and natural cycles in the form of solar irradiation, rain, wind, biogeochemical cycle and rivers (sources drawn on the left and up side of the diagram). These flows mostly supply local reservoirs (drawn in the form of storage) such as water, geothermal heat, soil and its organic contents, and materials for quarrying (*minerals*). Wastes are also conceived as a local stock of resources. Other inputs come from outside the system in the form of

---

[6] H.T. Odum, *Environment, Power and Society*, Wiley, New York, 1971; H.T. Odum, *Systems Ecology*, Wiley, New York, 1983 and H.T. Odum, *Environmental Accounting: Emergy and Environmental Decision Making*, Wiley, New York, 1996.

purchased goods, energy, services, materials and fuels (sources drawn on the up side of the diagram). Arrows are exchanges of energy and materials. Outputs exist in the form of outgoing arrows, for example products to the market, and the arrow to the bottom represents the *heat sink* that means dissipation, entropy production.

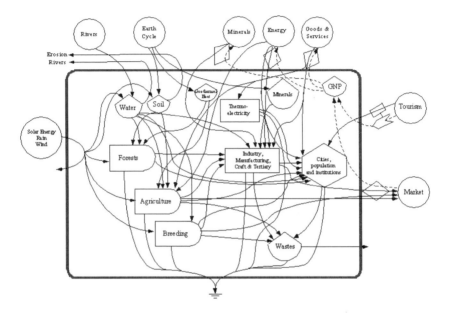

Figure 21: Energetic diagram of a generic urban region.

The *energy diagram* has been recently applied, by the *Ecodynamics group* of Siena University, to the Municipality of Ravenna, Italy, for a *Strategic Environmental Assessment* (SEA[7]) of the Master Plan (through an emergy synthesis). This study was aimed to account the main fluxes (exchanges of energy and matter) flowing through and within the boundaries of the system in order to direct strategic choices for territorial management taking into account the whole system behaviour and its dynamics in exploiting resources.

---

[7] The *Strategic Environmental Assessment* (SEA) is part of a Master Plan and is scheduled in the planning process during the preparation and before the plan is adopted, by national and regional Laws and according to the *European Council Directive 42/2001*.

7.2. Self-organization and cities

Entropy change depends on two processes: interactions among constituents of the system d_iS and exchanges with the external environment d_eS (see Fourth Step). How can we describe these processes in a complex urban system?

In a city or social system, the configuration of interactions between parts, such as individuals, and between the parts and the whole, is an expression of the system's organization. Single parts of the system interact with each other and organize themselves by exchanging information. This general organization extends to the whole system and is additional information that is not visible by observing the parts.

In complex systems, organization usually emerges spontaneously and is not imposed. The constituents do not decide the whole configuration with reference to a common project. On the contrary, the behaviour of the single elements is egoistic and individualistic, as Nicolis[8] pointed out, citing the law of *least resistance*: elements always modify their interactions with other elements, looking for the most advantageous. Interactions between system constituents (individuals), that are neither isolated nor free, involve continuous reciprocal adaptation; each individual acts and reacts according to actions and reactions of the other individuals.

This process, generated by competition and cooperation among the components, each pursuing its own aims, does not stop until organization that guarantees harmonious, non-conflicting interactions among individuals is achieved. Clear examples exist in nature, in which groups of agents, pursuing their own consistency, transcend themselves and acquire collective properties impossible to obtain individually. Organisms continuously adapt to each other by co-evolution until they constitute an ecosystem. Thought and life also belong to the category of complex living systems.

Societies, too, result from the aggregation of groups of individuals, and show emerging collective properties that single elements do not have. We know that certain individual behaviour is induced by society and how difficult it is to change a social behavioural trend or an economy, despite widespread awareness of the need to do so. In these cases, it is obvious that the whole is more than the sum of its parts.

The configuration of a complex system, or organization of parts into a whole, is therefore a steady state and depends on interactions among

---

[8] G. Nicolis, Physics of far-from-equilibrium systems and self-organization, in: P. Davies, ed., *The New Physics*, Cambridge University Press, New York, 1989.

constituents of the system (responsible for the entropy change d_iS) and perturbations from the external environment (responsible for the entropy change d_eS). The configuration is always dynamic: it is not static in time but changes whenever external conditions change. The main condition to maintain a state far from thermodynamic equilibrium, according to the concept of dissipative structures, is that the system must not be isolated and must exchange energy and matter with the external environment. This situation highlights the importance of interactions with the external environment; these also involve the constituents of the system for self-organization. Nicolis and Prigogine[9] provide some examples. A cubic centimetre of water at ambient temperature is characterized by the disordered thermal motion of its molecules; if it is subject to a winter storm, it self-organizes in the peculiar dendritic structure of a snowflake.

How do dynamics generate ordered structures in complex non-equilibrium systems? Why does system organization develop new configurations under certain conditions? External perturbations and conditions of nonlinearity may create and simultaneously destroy order, allowing new coherence to emerge beyond another bifurcation. Organization (or systems functioning without any apparent general law) depends on natural macroscopic processes generated by a multiplicity of unordered microscopic processes that, under certain conditions, depend on fluctuations. Processes of organization and self-organization in complex dissipative systems occur on different dimensional scales, and organization may emerge on a macroscopic space-time scale, many times bigger than the microscopic interactions between the elements[6].

Interactions are therefore what drives life and evolution. Self-organization depends on feedback from the external environment. The organization adopted by a system is the one that coordinates interactions between its constituents in a harmonious, non-conflicting way in the context. Small continuous fluctuations at the level of interactions between individuals can direct the choice of a whole configuration and even amplify the effect of a local perturbation until it pervades the whole system. This is based on the different directions of information exchange: information is received by the macro-structure and transmitted to the level of microscopic interactions between elements; information is also transmitted or amplified from the micro-level of individuals to the macro level of the whole system, due to nonlinearity.

---

[9] G. Nicolis and I. Prigogine, *Exploring Complexity. An Introduction*, Piper, Munich, 1987.

Interactions within a complex system, as well as those between the system and the outside, depend on fluctuations of internal or external origin that may generate new structures. The nature of the system is therefore adaptive and its evolution, drawn as a bifurcation diagram with many branches, depends on fluctuations. If a structural fluctuation due to a perturbation prevails, the whole system adopts a new mode of functioning (drawn as a new branch at a bifurcation point) with new syntax.

Dissipative structures, as complex self-adapting systems, can also be conceived as the result of fluctuations that, once established, can be maintained in a steady state with respect to a large set of perturbations. According to Prigogine, the emergence of spatial configurations or time rhythms in dissipative structures can be conceived as *order by fluctuation* and dissipative structures as *giant fluctuations* in a steady state. As stated by Prigogine[10]

> "A long way from equilibrium, the material acquires new properties, typical of non-equilibrium states, situations in which a system, far from being isolated, is subject to heavy external conditioning. And these completely new properties are truly necessary if we want to understand the world around us." [11]

7.3. Cities and mobile geographies

The perception of complexity in real systems suggested a new vision for cities. Researchers are currently seeking methods to visualize the dynamic evolutionary properties of urban systems in order to reveal their complex behaviour and the formation of ordered structures in a stationary state. Such research aims to recognize signs of complexity in urban systems and to understand how interactions between living and non-living systems generate collective properties and self-organizing processes.

Pattern formation by social activities, the dynamics of which affect urban texture, is the object of this research:

> "..., the manifestation of the powers that configure the city has shifted from the outwardly visible to the invisible – that is, the city is not rendered through composition, gravity, form, or material, as

---

[10] I. Prigogine and I. Stengers, *La Nouvelle Alliance*, Gallimard, Paris, 1979.

[11] I. Prigogine, *La Nascita del Tempo*, Bompiani, Milan, 1998.

much as it is through demographics and economic performance. [...]
no longer is the city visualised or composed as much as it is
empirically computed."[12]

By processing information through spatial (location) and temporal
parameters (dynamics), pattern formation and changing geographies may be
revealed in the urban context. The network of relations between society and
the urban environment, man and the city, living and nonliving entities, has an
adaptive evolutionary nature.

A joint research project, named *Mobile Landscapes*, was recently
conducted at the MIT SENSEable City Laboratory in Cambridge, USA, in
collaboration with the Department of Chemical and Biosystems Sciences
(*Ecodynamics group*) of Siena University, Italy, and in partnership with a
European telecommunications company. The project used techniques to
determine the geographic locations of cell phones in order to analyze
mobility patterns in towns. Location-based data from mobile devices was
collected and processed in an anonymous and aggregate form and plotted on
a sequence of maps. An immediate overview of mobile phone density in the
urban context was obtained.

Traffic data was related to the various antennas (base stations) that
recorded users connected to them for phone calls. This data is routinely
recorded by cell phone operators in the running of base station infrastructure
and is readily available. Data on the activity and location of the antennas can
be processed to show variations in daily (24 hours) or seasonal (long period)
intensity and its evolution in time. Based on national statistics for cell phone
use (almost everyone in Italy has a cell phone), the intensity of cell phone
activity is a likely indicator of population density.

The outcomes were in the form of mobile geographies revealing mobility
patterns based on real movements of people carrying cell phones in towns.
Computed maps of urban mobility therefore make it possible to visualize the
behaviour of people in urban areas with regard to the use of infrastructures,
buildings, services and other settings. Temporal modes of space use can be
appraised and revealed. This information is much more useful than any
conventional description of land use.

By revealing urban dynamics and time rhythms through a sequence of
configurations, by observing people's mobility and the emergence of critical
situations in the form of spatial nodes and temporary events and by showing

---

[12] R. Koolhaas, S. Boeri, S. Kwinter, N. Tazi and H.U. Obrist, *Mutations*, Actar, Art en Reve
Centre d'Architecture, Bordeaux, 2000.

the causes of congestion and unusual conditions, these maps make it possible to study the organization of urban systems and their self-adaptive dissipative nature. Besides potential applications for the management of practical problems and for urban planning, this method enables collective behaviours, perturbations and effects of fluctuations to be studied in complex urban systems with respect to our theoretical framework.

Regarding perturbations, for example, road-works or a new cafeteria with very good coffee can cause unexpected effects and significant variations in the pattern of urban activity, even recharging the organization of the entire neighbourhood. A new policy for traffic and work, road taxes, a new shopping mall, expansion of the wireless network, a new public transport line and many other events are examples of perturbations that make the system react and self-organize in a new pattern. Mobile Landscapes offered an opportunity to understand the changing complexity of contemporary cities. Its focus on temporal, rather than spatial patterns, is a new paradigm in urban analysis[13].

The first case study was the Milan metropolitan area. Location-based data, given in anonymous aggregate form, was collected around the clock for 16 days; it is in *erlangs*, a standard unit of traffic intensity in telecommunications systems (this unit combines the number of calls with the duration of calls; e.g., one erlang is the equivalent of a call one hour long, two calls half an hour long or 60 calls one minute long). Each traffic value is related to a cell, that is a circular area around a base station, covered by the GSM (Global System on Mobile Communications) signal. The study area was a square approximately 20 x 20 km, including 232 cells. Accuracy was around 400–800 metres. Traffic data was collected dynamically by the hour for each antenna. Location was fixed, namely the position of each base station.

A possible study would be to use this data to infer information about the "character" of a neighbourhood where an antenna is situated. Groups of cells can be classified on the basis of 'relative intensity' of cell phone activity during the 24 hour period, by observing different values in different time segments of the day; for example, districts with base stations showing prevalent use during working hours are likely to have an office/business vocation. Neighbourhoods with high evening and early morning cell phone traffic are likely to have a stronger residential character. Fig. 22 shows cells

[13] C. Ratti, R.M. Pulselli, S. Williams and D. Frenchman, Mobile landscapes: using location data from cell phones for urban analysis, *Environment and Planning B: Planning and Design*, in press, 2006.

with prevalence of activity in the evening and night (8 pm to 8 am) and during office hours (9 am to 1 pm and 2 pm to 6 pm). Associated with the residential or office "character" of a neighbourhood, this classification could be a powerful tool.

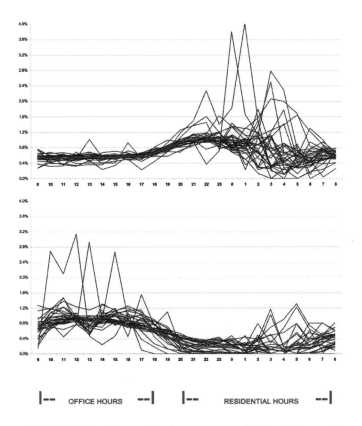

Figure 22: Mobile Landscapes Project: groups of cells with prevalence of activity in the evening and at night (residential areas) and during office hours (offices and services).

Different time bands can be analyzed during the 24 hour period through a sequence of maps. The results are like thermograms, highlighting the intensity of urban activity and its evolution in space and time. Fig. 23 shows a series over the study area between 8 am and 1 pm on 19th April 2004. It shows that the relative intensity of activity was maximum in the suburbs early in the morning, moving into the city centre (mostly offices) during the morning, and peaking in the central district at lunchtime.

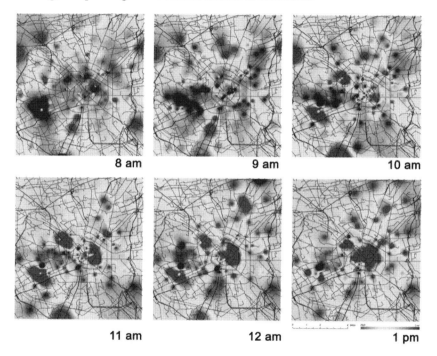

Figure 23: Mobile Landscapes Project: maps of Milan from 8am to 1pm[13].

This pilot project in Milan enabled all steps of the technique to be tested (selection of data-base, data collection, data processing, plotting on maps, interpretation of results). The method will be applied by increasing its accuracy with a data set having high detail in terms of time segments and spatial accuracy.

Location-based services recently raised a controversy about privacy. This project is based on anonymous aggregate data and it certainly does not seek to spy on individuals. Data management is irreprehensible and helps to suggest rules on data management for emerging cases. Far from spying on people's behaviour or suggesting any *big brother* policy, the aim of this project is to use location based services for the common good.

The subject of study is the life of human systems, which is why mobile geographies work at the scale of huge urban areas and focus on the evolution of the whole social system with its emerging collective properties. The behaviour of the system, consisting of many interactions between single elements, is studied as a whole; the configuration is a network of relations that can change with respect to a series of local perturbations. With regard to evolution and design, Erich Jantsch writes:

> "Human life is movement. It is movement not by and for itself, but within a dynamic world, within movements of higher order. [...] these higher-order movements constitute the life of human systems."[14]

7.4. Cities and design

Evolutionary thermodynamics and complex system theory provide a potential scientific background for many other fields of application. The paradigm shift suggested by Prigogine is an opportunity to ask new questions and define an approach for understanding nature, cities, society and life. Knowledge of nature must come from a global, systemic view of events and from study of the network of information joining the various forms of matter, energy and life in time and space[15]. The perception of complexity in networks and dissipation in far-from equilibrium systems gives a theoretical framework that also works for reloading principles of aesthetics.

This engaging theory has its potential playground in contemporary cities. Urban studies and design have to focus on networks of relations spread over multiple textures. The urban framework, apparently static and inert, is the stage for complex dynamics, interactions, exchanges and emerging phenomena that cause multiple experiences, sensations and emotions. Organization in networks, not always continuous and linear, distributed in

---

[14] E. Jantsch, *Design for Evolution: Self-organization and Planning in the Life of Human Systems*, George Braziller, New York, 1975.
[15] E. Tiezzi, *Beauty and Science*, WIT Press, Southampton, UK, 2004.

space with non-homogeneous intensity and changing coverage is a key for interpreting towns, conceived as nodes in which connections increase and achieve high intensities. Urban space, a node, changes its sense/function according to network intensity and network dynamics. When the network changes, the sense of place changes as well.

This idea of city as a place of interactions with adaptive dissipative properties has suggested unusual experiments in the field of urban design and architecture. Contemporary architecture has come up with surprising urban shapes related to the formation of connections, networks, over-layered textures and patterns. The complexity of today's cities has raised much discussion on town planning and management. "Without any intermediation, architectural language has appropriated the dimension of the city in its fullest extent, turning itself into a totally urban language", as Purini[16] underlines:

"layouts and fabrics, spaces and places, settings and contexts have been adopted by the construction as its genetic memory, a hidden substructure that invisibly nourishes composition".

Architects such as Zaha Hadid, Frank Gehry, Peter Eisenman, Daniel Libeskind and Enric Miralles, probably aware of the work of Prigogine, Lorenz and Mandelbrot, have developed intuitive, provocative urban metaphors. Buildings designed by these architects simulate self-organizing dissipative structures, fractal geometries and strange attractors, being framed in an arbitrary stochastic way.

Hadid's concept for the MAXXI – *XXI Century Arts National Museum* – in Rome (Fig. 24) was a node, an intersection of potential flows of relations from the immediate neighbourhood and remote places, that cross, overlap and interlace to each other. Multiple connections with the outside are frozen in an astonishing form, a self-organizing structure that simulates, as a metaphor, the stationary state of a system far from thermodynamic equilibrium. The introverted deconstructive shape of Gehry's Guggenheim museum in Bilbao (Fig. 25) was conceived as a collision of volumes and unusual spaces made by distortion of convex surfaces, framed and interrupted, and gravitating around a virtual orbit similar to a strange attractor.

---

[16] F. Purini, News awaited for some time, *Lotus*, **104**, 60–67, Electa, Milano, 2000.

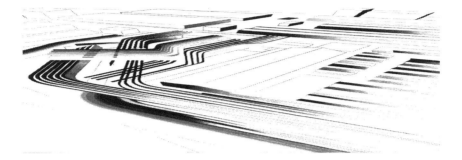

Figure 24: MAXXI – XXI Century Arts National Museum in Rome –
architect Zaha Hadid (2000).

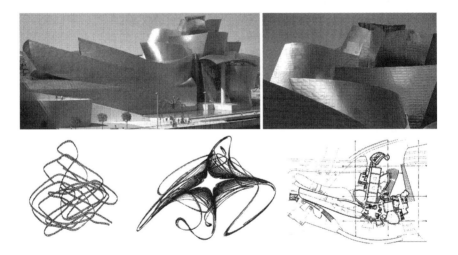

Figure 25: The Guggenheim Museum in Bilbao – architect Frank Gehry
(1997) – and strange attractors.

The project of the Virtual Guggenheim museum by Asymptote Studio
(Fig. 26) is a kind of Moebius strip in cyberspace that resembles Lorenz's
strange attractor.

Figure 26: Lorenz's strange attractor and the Virtual Guggenheim Museum by Asymptote Studio (1999).

Architecture is no longer conceived as an inorganic box, a nonliving support. The demand for specific requisites such as comfort, aesthetics and function in different environmental contexts suggests a much more comprehensive interpretation. Interactions between buildings and living organisms, people and the environment, require exchange of energy and matter to achieve and maintain a stationary state, as do living organisms. These intuitions of contemporary architecture, even in this ephemeral form, enhance the evolutionary nature of cities or even show the inadequacy of traditional town planning and architecture with respect to the concept of complexity and self-organization in an urban context.

The challenge and novelty of contemporary architecture lies in considering the dynamics of urban systems, flows of energy and matter, information exchange and the formation of ordered structures. The task of the next generation of architects, aware of the complexity of the dissipative contemporary city and systems far from thermodynamic equilibrium, is to transform the provocative suggestions of the turn of the century into coherent architecture.

7.5. The anti-aesthetic assumption

We again underline, as did Gregory Bateson[17,18] the anti-aesthetic dogma according to which quality and beauty do not have scientific dignity. Do you remember the tyranny of the clock in *Metropolis* and the robot built for this Fritz Lang film of 1926? In some ways, *Metropolis* was an ecological cult

---

[17] G. Bateson, *Steps to an Ecology of Mind*, Chandler, San Francisco, 1972.
[18] G. Bateson, *Mind and Nature: a Necessary Unity*, Dutton, New York, 1979.

movie before its time. Those were times when technological advances still produced the illusion of "progress", though military applications of machines should already have sounded a warning. They were the times of the first catastrophes: the Hindenburg, the Titanic. Breakdowns of the proportions of Chernobyl or New Orleans were unimaginable, but the apprentice sorcerers of reductionism and technocracy were already at work, reducing ancient arts and creative wisdom to machines. Space squeezed time out of scientific thought, reducing it to a reversible interval. Numbers and quantities swept away creativity and quality.

Once, journeys to Italy were a must for any true European intellectual, and rich records of these travels go from G. Berkeley to Goethe. Today, "travellers" can no longer recognize the city they are approaching or tell the transition from the Florentine to the Sienese countryside indicated by the transition from *pietra serena* to *travertino* in the architraves of houses. The tourist of today cannot tell whether he is in Bergamo or Lecce from the materials used in the outskirts of these towns: sandstone and tuff are no longer milestones of journeys.

There was an Arabian geographer at the court of the Norman king in Palermo. They say that Frederick II held Idris in high esteem, and Idris returned this esteem with maps of lands and seas from Sicily to northern Italy. There is a wonderful map with a drawing of the town of Lucca. Idris showed routes of ships and tracks for horses. The maps had one peculiarity: the distances were not proportional to space but to travelling time. These temporal maps not only plotted space but also time. Clearly it takes longer to cross mountains or a forest than a plain.

In construction materials too, we can no longer read the passing of time. Buildings are like wine: the older they are, the better they become, if they are well constructed. Some wines turn to vinegar and some buildings age badly. When brick and stone challenge time, "harmony beats silence by a thousand centuries"[19].

---

[19] U. Foscolo, *I Sepolcri*, 1807.

8
EIGHTH STEP

ORIENTORS, GOAL FUNCTIONS AND CONFIGURATIONS OF PROCESSES

Blue as the sky, the sea,
oh mountains! oh rivers! it was better to stay,
not to look beyond, to dream:

dream is the infinite shadow of Truth [...]

And thus he weeps on conquering:
the brown eye weeps like death;
the blue eye weeps like the sky.

He is ever thus (that is his fate)
in the brown eye, vain hope;
in the blue eye, ardent desire.

Giovanni Pascoli[1]

---

[1] G. Pascoli, *Poesie*, Mondadori, Milano, 1944. "Alexandros": "Azzurri come il cielo, come il mare, o monti! o fiumi! era miglior pensiero ristare, non guardare oltre, sognare: il sogno è l'infinita ombra del Vero. [...] E così, piange, poi che giunse anelo: piange dall'occhio nero come morte; piange dall'occhio azzurro come cielo. Ché si fa sempre (tale è la sua sorte) nell'occhio nero lo sperar, più vano; nell'occhio azzurro il desiar, più forte".

8.1. Linger fair passing moment

The verse in Goethe's Faust and the poem of Pascoli describe tension between desire to know everything immediately and the wisdom of stopping to reflect and contemplate.

The tension described by Pascoli in the poem dedicated to Alexander the Great is between desire to push ahead and the disappointment experienced on reaching the destination. "We have arrived: it is the End", says Alexander to his soldiers, and indicating the moon, adds, "no other Land but there, in the air".

Verses dedicated to nature, to the mountains and rivers, tell a different story, a story of dreams and beauty, a story that says, "Linger, fair passing moment". This, too, is a story of tension, but harmonic tension between two desires: desire to press ahead and desire to stop, contemplate, dream.

We find the same tension in the verses of Antonio Machado: "Wanderer, your footprints are the way and nothing else; Wanderer, there is no way, the way is made by going [...] Wanderer, there is no way but wakes in the sea".

From chemical and physical points of view the Universe consists of matter and energy. However, in the Universe there is something more. There are beauty and biodiversity, there is free will and the possibility of choice, there are preferences, as outlined by the master of "ecology of mind", Gregory Bateson[2]:

> "We know something about Nature's preferences. She prefers the probable to the improbable and if she were guided only by this preference, known as the second law of thermodynamics, the Universe would be simple, though somewhat tedious. But Nature has another evident preference: she prefers stability to instability. This preference, of itself, would also lead to a tedious Universe. It is the combination of these two preferences that makes the Universe in which we live extremely complex and strangely unpredictable. In an Universe based only on probability or stability, there would be neither surprises nor evolution; there would be no organisms capable of being surprised.
>
> The fantastic mobile game that all organisms play, or rather, all particles of the Universe, depends on this double system of preferences that seems to be characteristic of Nature."

---

[2] G. Bateson, *A Sacred Unity – Further Steps to an Ecology of Mind*, The Estate of Gregory Bateson, USA, 1991.

These are the blue and the brown eyes, this is the game to play for an evolutionary science.

In order to understand the evolutionary properties of the Universe and of the biosphere we must shift our attention to structural organization and the information embodied in matter and energy. To do this, a gestalt shift is necessary from conservative and state functions to trend and goal functions.

8.2. From state functions to goal functions (orientors)

This step is dedicated to the footprints leading to evolutionary physics, to new goal functions (or orientors) for ecosystems and living systems.

To understand the state and dynamics of systems with high complexity and synergy (such as living systems and ecosystems) new scientific tools are emerging. The functions used to investigate the state and dynamics of these systems are called *goal functions or (ecological) orientors*[3].

Systems characterized by self-organizing processes and dissipative structures, that build gradients and order from thermodynamic equilibrium and microscopic disorder, seem to have common behaviours. Certain collective features emerge and similar attributes can be observed, even between very different environments. The development of these systems seems to be oriented towards specific areas in state space (*attractors*).

Such behaviour is frequent in living systems and ecosystems, which also show enormous creativity in their evolutionary paths. In spite of the wide variability of choices typical of natural systems, such oriented trends are strongly present.

Living systems usually adapt to changes coming from the environment, tending to *maintain themselves as far as possible from thermodynamic equilibrium* by creating gradients in temperature, pressure, chemical composition, structures, etc. Certain extreme principles, that can be observed in nature, direct the evolution of living systems and ecosystems. They have been used as references to formulate all known goal functions.

As we shall see, several different orientors have been proposed and acknowledged by the scientific community in the last 20 years. Each has its own features and is based on different observed behaviours of natural systems, but all have some general properties in common.

---

[3] F. Müller and M. Leupelt, eds., *Eco Targets, Goal Functions, and Orientors*, Springer-Verlag, Berlin, 1998.

Goal functions must describe properties of ecosystems or living systems that are regularly optimized during evolution. This is a necessary feature of goal functions.

An ecological orientor should also characterize system state and distinguish different systems. This means that a goal function can be used to classify the state of a system from a developmental point of view and should indicate its degree of maturity. This property can be used to compare and classify systems.

Goal functions describe systems from a holistic point of view and define their structural and functional features. They are based on certain general principles, such as thermodynamic principles, and they have to reflect the general properties of living dissipative self-organizing systems. Their ability to measure self-organization capacity suggests that ecological orientors can also indicate some aspects of the degree of naturalness of ecosystems. They can therefore be used to evaluate the strength of human impact and an ecosystem's structural carrying capacity. Goal functions can therefore provide "a good basis for finding usable indicators for ecosystem health, ecological integrity and sustainability"[3].

The following sections give a classification of existing orientors and a general description of those most widely used.

8.3. Classification of goal functions

Before describing goal functions, let us delineate the general features of dissipative (living) self-organizing systems[4]:

(1) it is necessary that the system be open (or at least non-isolated) to exchange energy (as well as mass) with its environment;

(2) an influx of low-entropy energy that can do work is necessary;

(3) an outflow of high-entropy energy (heat produced by transformation of work to heat) is necessary (this means that the temperature of the system must inevitably be greater than 2.726 K);

(4) entropy production accompanying the transformation of energy (work) to heat in the system is a necessary cost of maintaining the order;

---

[4] S.E. Jørgensen and Y. M. Svirezhev, *Towards a Thermodynamic Theory for Ecological Systems*, Elsevier, Oxford, UK, 2004.

(5)   mass transport processes at a not too low rate are necessary (a prerequisite);

(6)   abundant presence of the unique solvent water is a prerequisite for the formation of life forms similar to the life forms as we know from Earth;

(7)   the presence of nitrogen, phosphorus, and sulphur and some metal ions seems absolutely necessary for the formation of carbon-based life;

(8)   as the formation of life from inorganic matter requires a very long time, probably of the order of 10^8 years or more, the seven conditions have to be maintained in the right ranges for a very long time, which probably exceeds about 10^8 years.

Orientors have four "germinal" points in the partially overlapping fields of Lotka's *Maximum Power Principle*[5], information theory (see e.g., Shannon and Weaver[6]), graphs and networks[7,8] and dissipative structures[9]: different mixing of these "ingredients" has produced a number of goal functions, which have several similar and other complementary aspects (for a complete survey of the characteristics of orientors, see Fath et al.[10]).

Systems can be treated and analyzed statically using the well-established techniques of graphs and information theory. Alternatively, they may be viewed dynamically as dissipative structures that exist by virtue of energy flows and of building energy into structure. This entails changes in the transition from high quality, low-entropy forms to lower quality, intermediate-entropy forms, finally ending up as the highest-entropy form, heat. In this case, thermodynamic efficiencies can be examined using methods for optimized functioning as suggested in the current literature[9].

---

[5] A.J. Lotka, Contribution to the energetics of evolution, *Proceedings of National Academy of Sciences*, **8**, 147–151, USA, 1922.

[6] C.E. Shannon and W. Weaver, *The Mathematical Theory of Communication*, University of Illinois Press, USA, 1949.

[7] B.C. Patten and E.P. Odum, The cybernetic nature of ecosystems, *Amer. Nat.*, **118**, 886–895, 1981.

[8] M. Higashi, B.C. Patten and T.P. Burns, Network trophic dynamics: the modes of energy utilization in ecosystems, *Ecological Modelling*, **66**, 1–42, 1993.

[9] S.N. Nielsen and R.E. Ulanowicz, On the consistency between thermodynamical and network approaches to ecosystems, *Ecological Modelling*, **132**, 23–31, 2000.

[10] B.D. Fath, B.C. Patten and J.S. Choi, Complementarity of ecological goal functions, *J. Theor. Biol.*, **208**, 493–506, 2001 and B.D. Fath, A nonthermodynamic constraint to trophic transfer efficiency based on network utility analysis, *Int. J. Ecodynamics*, **1**, 28–43, 2006.

Thermodynamic-based goal functions describe how self-organized ecological systems develop by keeping their state as far as possible from thermodynamic equilibrium. *Exergy, entropy, emergy* (power), exergy-empower ratio and *ascendency* are widely used thermodynamic orientors.

Energy is considered a sort of *a priori* concept in thermodynamics, but energy manifests in several forms, each endowed with a different *quality*. The concept of quality can only be explained through the Second Law. "High-quality" energy forms (potential, kinetic, mechanical, electrical, etc.) can theoretically be converted into each other with 100% efficiency. Other forms (internal energy, chemical energy, thermal radiation, turbulent kinetic energy, etc.) cannot be converted into each other efficiently, and much less into high-quality energy. The concept of exergy was developed to provide a congruent and coherent quantification of the quality of form of energy.

Exergy is defined as the maximum work developed in an ideal process. Note that since the work arises from an interaction between the system and the environment, exergy is an attribute of the pair (system, environment), and not of the system alone. The system and the environment exchange energy flows in different forms: thermal (heat flows from the system to the environment or *vice versa*), work (the system performs or receives work from its boundary interactions with the environment), mass exchange (mass flows from the system to the environment and *vice versa*), etc.

As exergy is the amount of work that a system can perform when it is brought into chemical equilibrium with its environment, exergy can also be seen as a measure of the distance of the system from thermodynamic equilibrium, or in biological terms, the condition of death.

The major contributions for the creation of ecological orientors based on exergy have been given by Jørgensen with the concept of eco-exergy[11,12] and by Schneider and Kay[13].

To apply exergy to living systems, especially ecosystems, it is necessary to distinguish between different levels of complexity: for example, the elementary composition of 1kg of fish and 1kg of phytoplankton can be considered very similar, but in fact their complexity is quite different. Jørgensen and coworkers have introduced "weights" to distinguish levels

---

[11] S.E. Jørgensen and H.F. Mejer, A holistic approach to ecological modelling, *Ecological Modelling*, **7**, 169–189, 1979.

[12] S.E. Jørgensen, Exergy and ecology, *Ecological Modelling*, **63**, 185–214, 1992.

[13] E.D. Schneider and J.J. Kay, Life as a manifestation of the second law of thermodynamics, *Mathematical Computer Modelling*, **19**, 25–48, 1994.

with different genetic complexity[14]. Eco-exergy becomes a weighted sum of concentrations in which the different "weights" are functions of the number of genes of each species. Eco-exergy is not an absolute indicator because of several approximations, and becomes a relative indicator, very useful for comparing different ecosystems or monitoring an ecosystem during time. The exergy input, which brings highly usable energy, needs to be degraded as much as possible by the ecosystem and converted into structural exergy that remains in the ecosystem in the form of more complex organisms and structures. Indeed, ecosystems seem to maximize both eco-exergy storage[15,16] and the capture of external exergy flows[17].

The work by Prigogine is the reference for thermodynamic orientors based on entropy functions. The first orientor is represented by maximization of total entropy production[18,19] that corresponds to maximization of the dissipation of input flows according to Kay's exergy degradation principle, already mentioned[13].

The second orientor is based on the minimum specific dissipation principle[20,21,22] that demonstrates that the organization of an ecosystem tends to maximize the efficiency of internal energy transfers and conversions through internal feedbacks and to maximize the residence time of input flows. In this context, Aoki's[23] application of entropy production and flow to

---

[14] J.C. Fonseca, J.C. Marques, A.A. Paiva, A.M. Freitas, V.M.C. Madeira and S.E. Jørgensen, Nuclear DNA in the determination of weighting factors to estimate exergy from organisms biomass, *Ecological Modelling*, **126**, 179–189, 2000.

[15] G. Bendoricchio and S.E. Jørgensen, Exergy as a goal function of ecosystems dynamics, *Ecological Modelling*, **102**, 5–15, 1997.

[16] S.E. Jørgensen, J. Marques and S.N. Nielsen, Structural changes in an estuary, described by models and using exergy as orientor, *Ecological Modelling*, **158**, 233–240, 2002.

[17] E.D. Schneider and J.J. Kay, Complexity and Thermodynamics. Toward a new ecology, *Futures*, **26**, 626–647, 1994.

[18] I. Prigogine and I. Stengers, *Order Out of Chaos: Man's New Dialogue with Nature*, Bantam Books, New York, 1984.

[19] D.R. Brooks and E.O. Wiley, *Evolution as Entropy: Toward a Unified Theory of Biology*, University of Chicago Press, Chicago, 1986.

[20] L. Onsager, Reciprocal relations in irreversible processes, *Phys. Rev.*, **37**, 405–426, 1931.

[21] L. Onsager, Reciprocal relations in irreversible processes, *Phys. Rev.*, **38**, 2265–2279, 1931.

[22] I. Prigogine, *Thermodynamics of Irreversible Processes*, Wiley, New York, 1955.

[23] I. Aoki, Entropy and exergy in the development of living systems: a case study of lake-ecosystems, *J. Phys. Soc. Japan*, **67**, 2132–2139, 1998.

living systems and ecosystems is very interesting, as is use of the respiration/biomass ratio in lacustrine communities by Choi et al.[24].

Odum[25,26] started from the work of Lotka and his Maximum Power Principle, adding a concept of quality of energy that is different from that used in exergy. Odum's "quality" is connected with how much is necessary to obtain something, instead of how much is obtainable from something.

"Emergy is the available energy of one kind that was used directly and indirectly to make a service or product"[27]. In particular, *solar emergy* (measured in solar emergy Joules – sej) is the solar energy required (directly or indirectly) to make a service or product. (Solar) Emergy is a measure of the energy contributed to a product by work done by global environmental and human services in space and time. It is sometimes referred to as "energy memory"[28] and its logic ("memorization" rather than "conservation") is different from other energy-based analyzes, as shown by emergy "algebra"[29].

The total emergy flowing through a system per unit of time is called *empower*.

According to Lotka's principle, each transformation surviving self-organization helps maximize its power and reinforces the network[30]. Odum referred to Lotka's principle as "ambiguous", since it seems to imply that transformations at a lower scale are more important than those at a higher scale, since they carry more energy[27]. Indeed, 1 MJ in the form of a herbivore is very different from 1 MJ in the form of grass: the herbivore implies about 10 MJ of energy in the form of grass[31]. This is why Odum modified Lotka and stated that "if natural selection is given time to operate, the higher the emergy flux necessary to sustain a system or a process, the higher its

[24] J.S. Choi, A. Mazumder and R.I.C. Hansell, Measuring perturbation in a complicated, thermodynamic world, *Ecological Modelling*, **117**, 143–158, 1999.

[25] H.T. Odum, Self-organization, transformity and information, *Science*, **242**, 1132–1139, 1988.

[26] H.T. Odum, in: C. Rossi and E. Tiezzi, eds., *Ecological Physical Chemistry*, Elsevier, Amsterdam, 25–65, 1991.

[27] H.T. Odum, *Environmental Accounting: Emergy and Environmental Decision Making*, Wiley, New York, 1996.

[28] D. Scienceman, Energy and emergy, in: G. Pillet and T. Murota, eds., *Environmental Economics*, Leimgruber, Geneve, Switzerland, 257–276, 1987.

[29] M.T. Brown and R.A. Herendeen, Embodied energy analysis and Emergy analysis: a comparative view, *Ecological Economics*, **19**, 219–235, 1996.

[30] S.E. Jørgensen, Toward a thermodynamics of biological systems, *Int. J. Ecodynamics*, **1**, 9–27, 2006.

[31] S. Bastianoni, Emergy, empower and the eco-exergy to empower ratio: a reconciliation of H.T. Odum with Prigogine?, *Int. J. Ecodynamics*, in press, 2006.

hierarchical level and the usefulness that can be expected from it"[24] or "prevailing systems are those whose designs maximize empower by reinforcing resource intake at the optimum efficiency"[27].

This is in line with maximization of dissipation or entropy production (Schneider and Kay[13], Odum[27], Fath[10]), but seems at variance with Prigogine's theorem of minimum entropy production[22].

Bastianoni[32,33] proposed combined use of eco-exergy and empower. According to Fath et al.[10], minimization of the empower/stored exergy ratio also seems to be a primary goal function in natural systems.

The ratio of emergy flow to eco-exergy indicates the solar emergy flow required by an ecosystem to produce or maintain a unit of complex system organization or structure (Bastianoni and Marchettini)[33]. We actually find the eco-exergy to empower ratio more meaningful since it reflects the state of the system (as eco-exergy) per unit input (as emergy flow). The eco-exergy/empower ratio can be regarded as the efficiency of an ecosystem in transforming direct and indirect inputs of basic energy (solar) into organization of the living system or ecosystem. If the eco-exergy/empower ratio has an increasing trend (apart from oscillations due to normal biological cycles), it means that natural selection is taking the system along a thermodynamic path that will bring it to a higher organizational level.

In the last few years, R.E. Ulanowicz developed a new interpretation of ecosystems dynamics based on *ascendency*[34,35,36]. Linking typical concepts of thermodynamics with information theory, Ulanowicz provides a view that makes it possible to distinguish between growth and development of an ecosystem. Ulanowicz gives both qualitative and quantitative description of the interactions of ecosystem components. The changes caused by these interactions within the ecosystem do not obey strictly deterministic laws but have a strong stochastic component. Thus probability and information theory have a very important role in the analysis of these systems. Ulanowicz defined ascendency as the product of two components: the information

---

[32] S. Bastianoni, A definition of pollution based on thermodynamic goal functions, *Ecological Modelling*, **113**, 163–166, 1998.

[33] S. Bastianoni and N. Marchettini, Emergy/exergy ratio as a measure of the level of organization of systems, *Ecological Modelling*, **99**, 33–40, 1997.

[34] R.E. Ulanowicz, *Growth and Development, Ecosystems Phenomenology*, Springer-Verlag, New York, 1986.

[35] R.E. Ulanowicz, *Ecology, The Ascendent Perspective*, Columbia University Press, New York, 1997.

[36] R.E. Ulanowicz, Process Ecology: a transactional worldview, Moen Brainstorming Meeting and *Int. J. Ecodynamics,* in press, 2006.

contained in the ecological network, called Average Mutual Information (AMI), and Total System Throughput (TST). The former is related to the development of the system, the latter to its growth. Ecological successions often occur by increasing ascendency. An increase in the number of species, greater resource use within the system and increasing trophic specialization are reflected as an increase in ascendency. A system without major perturbations therefore tends to have increasing ascendency. The maximum ascendency principle was formulated by Ulanowicz as a condition towards which ecosystems naturally tend.

Ulanowicz[36] introduces the concept of *configurations of processes* (see Eleventh Step).

Each of the above goal functions is valid in some aspect of the analysis of ecosystems. They show viewpoints that may be partly independent but also consistent with each other. Fath et al.[10] showed that ten extremal principles involving orientors (*maximum power, maximum storage, maximum empower, maximum emergy, maximum ascendency, maximum dissipation, maximum cycling, maximum residence time, minimum specific dissipation,* and *minimum empower/exergy ratio*) can be unified by ecological network notation. Table 2 (from Fath et al.[10]) shows how the different orientors can be expressed in terms of network parameters according to their principles. Extremal principles and the associated network formulations are given, where f indicates a flow, x a storage and τ the turnover time. Superscripts denote modes. As these are system-wide properties, the appropriate notation is used: TST = total system throughflow, TSS = total system storage, TSE = total system export, TSC = total system cycling, EMP = empower, EMG = emergy, ASC = ascendency, AMI = average mutual information.

Table 2: Goal functions and their network analysis formulation[10].

Principle	Extremal principle	Network parameter (system level)	Network analysis formulation
Maximize power	max{TST}	$TST = f^{(0)} + f^{(1)} + f^{(2)}$	$TST = \sum\sum (n_{ij}) z_j$
Maximize storage	max{TSS}	$TSS = x^{(0)} + x^{(1)} + x^{(2)}$	$TSS = \sum\sum \tau_i(n_{ij}) z_j$
Maximize empower	max{EMP}	$EMP = f^{(0)} + f^{(1)} + f^{(2)}$	$EMP = \sum\sum (n_{ij}^a) z_j$
Maximize emergy	max{EMG}	$EMG = x^{(0)} + x^{(1)} + x^{(2)}$	$EMG = \sum\sum \tau_i(n_{ij}^a) z_j$
Maximize ascendency	max{ASC}	$ASC = AMI^*[f^{(0)} + f^{(1)} + f^{(2)}]$	$ASC = AMI^* \sum\sum (n_{ij}) z_j$
Maximize dissipation	max{TSE}	$TSE = f^{(4)}$	$TSE = \sum\sum \varepsilon_i(n_{ij}) z_j$
Maximize cycling	max{TSC}	$TSC = f^{(2)}$	$TSC = \sum\sum (n_{ij}/n_{ii})(n_{ii} - 1) z_j$
Maximize residence time	max{TSS/TST}	$TSS/TST = \tau$	$TSS/TST = \sum\sum \tau_i(n_{ij}) z_j / (n_{ij}) z_j = \sum \tau_i$
Minimize specific dissipation	min{TSE/TSS}	$TSE/TSS = f^{(4)}/(x^{(0)} + x^{(1)} + x^{(2)})$	$TSE/TSS = \sum\sum \varepsilon_i(n_{ij}) z_j / \tau_i(n_{ij}) z_j$ $= \sum\sum \varepsilon_i / \tau_i$
Minimize empower to exergy ratio	min{TST/TSS}	$TST/TSS = 1/\tau$	$TSS/TST = \sum\sum (n_{ij}^a) z_j / \tau_i(n_{ij}) z_j$ $= \sum\sum (n_{ij}^a) / \tau_i(n_{ij})$

Fath et al. also try to formulate a general principle encompassing all aspects of the orientors: "Get as much as you can (maximize input and first-passage flow), hold on to it for as long as you can (maximize retention time), and if you must let it go, then try to get it back (maximize cycling)"[10]. They conclude that the *résumé* of these three aspects lies in minimization of specific dissipation, as stated by Prigogine[22].

The two principles that seem most contradictory are *maximize dissipation* and *minimize specific dissipation*, however both can co-occur if total system storage increases faster than total system export. Minimizing specific dissipation combines output (and by equivalence, input) and storage into one organizing principle such that both dissipation and structure are maximizing while at the same time their ratio is minimizing. The same holds for the minimization of the empower/exergy ratio, which can be seen as a reconciliation between Odum and Prigogine[31].

Although each of the proposed goal functions describes different aspects of systems dynamics well, the best way to exploit their complementarities and interdependencies is to use a multiple approach.

9
NINTH STEP

SEMANTIC BIOPHYSICAL CHEMISTRY

Now come, and next hereafter apprehend
What sorts, how vastly different in form,
How varied in multitudinous shapes they are-
These old beginnings of the universe;
Not in the sense that only few are furnished
With one like form, but rather not at all
In general have they likeness each with each,
No marvel: since the stock of them's so great
That there's no end (as I have taught) nor sum,
They must indeed not one and all be marked
By equal outline and by shape the same.

Moreover, humankind, and the mute flocks
Of scaly creatures swimming in the streams,
And joyous herds around, and all the wild,
And all the breeds of birds – both those that teem
In gladsome regions of the water-haunts,
About the river-banks and springs and pools,
And those that throng, flitting from tree to tree,
Through trackless woods – Go, take which one thou wilt,
In any kind: thou wilt discover still
Each from the other still unlike in shape.(...)

> *(...)Lastly, with any grain,*
> *Thou'lt see that no one kernel in one kind*
> *Is so far like another, that there still*
> *Is not in shapes some difference running through.*
> *By a like law we see how earth is pied*
> *With shells and conchs, where, with soft waves, the sea*
> *Beats on the thirsty sands of curving shores.*
>
> Lucretius[1]

9.1. The elm table

This beautiful elm table is made of the same material as an elm tree but from the point of view of natural and chemico-physical behaviour, the table and the tree are completely different, not only because the tree grows and reproduces but also because it embodies the information and capacity of self-organization that the elm table lacks. These properties are related to Prigogine's *dissipative structures*, to Varela's *autopoiesis* and to *epigenesis*. The tree is a thermodynamic system far from equilibrium that maintains itself in life; the elm table can only eventually reach thermodynamic equilibrium, obeying the laws of classical physical chemistry. The properties of the elm table depend on the sum of the molecules that compose it.

The elm tree will also reach thermodynamic equilibrium but only after it goes from the living to the nonliving state. The sciences that study living things are different from those concerned with nonliving things, both in

---

[1] Lucretius, *De rerum natura*, **II**, 335–375, "Nunc age, iam deinceps cunctarum exordia rerum qualia sint et quam longe distantia formis, percipe, multigenis quam sint variata figuris; non quo multa parum simili sint praedita forma, sed quia non volgo paria omnibus omnia constant. nec mirum; nam cum sit eorum copia tanta, ut neque finis, uti docui, neque summa sit ulla, debent ni mirum non omnibus omnia prorsum esse pari filo similique adfecta figura. Praeterea genus humanum mutaeque natantes squamigerum pecudes et laeta armenta feraeque et variae volucres, laetantia quae loca aquarum concelebrant circum ripas fontisque lacusque, et quae pervolgant nemora avia pervolitantes, quorum unum quidvis generatim sumere perge; invenies tamen inter se differre figuris.(...) (...)Postremo quodvis frumentum non tamen omne quidque suo genere inter se simile esse videbis, quin intercurrat quaedam distantia formis. concharumque genus parili ratione videmus pingere telluris gremium, qua mollibus undis litoris incurvi bibulam pavit aequor harenam.".

epistemological and logic properties (Pascal versus Descartes), complexity and mathematics (e.g. goal functions versus state functions).

The law of Pascal clearly holds for the elm tree which is more than the sum of its parts. The difference between the elm table and the elm tree lies in the role of the information contained in the elm tree and its signification (semantics).

Here it is worth pointing out that although reproduction is essential for the continuation of the species, it is not essential for life. Humans, trees and animals can continue to live without reproducing and they can create wonderful things (art, science, forms, colours, sounds, scents, etc.). We shall discuss the roles of metabolism and reproduction and their relationships with physical chemistry, especially entropy. The question implied by these things is: *What is life?*

9.2. The apricot paradox

The palaeontologist Roberto Fondi[2] introduces an important step to define the watershed between living and not living systems. Fondi asks:

> "How does a living cell differ from other physical systems? There have been many answers to this question, the most appropriate of which seems to be: *cells differ from other physical systems by virtue of the increased complexity inherent in their epigenetic development,* or in other words due to a series of geneses, each of which creates new structures and new functions. No machine or inanimate physical system can increase its complexity as can the simplest living cell.
>
> A milestone in twentieth century science was the discovery that just as the information of a word is determined by the linear order of its letters, the information of a gene, DNA, is determined by the linear order of its nucleotides, the triplets of which "recall" specific amino acids for the building of proteins. Protein function depends only on the three-dimensional disposition of its amino acids. Since genes obviously cannot transport the information necessary for building proteins into this space (*as if writing "apricot" on a strip of paper would enable the paper to roll up and become a real apricot*), the enormous quantity of information necessary for that construction

---

[2] R. Fondi, personal communication and *Int. J. Ecodynamics*, submitted.

can only come from *the increase in complexity inherent in an epigenetic process.*

As shown by Marcello Barbieri[3], a physical system can increase in complexity only if it has memory and a translation code. Like the genetic codes present in all organisms, the linguistic codes exclusive to our species, and many other codes yet to be deciphered, memories and translation codes *must therefore be fundamental components of all organisms.*

This conclusion radically changes our way of looking at the living world, making it impossible to interpret the world as a mere assemblage of objects dominated by the rigid deterministic dialectic of chance and necessity, but confers the world with properties hitherto considered exclusive to the mind or psyche. It is therefore impossible to explain the origin of cell systems without simultaneously explaining the origin of the memories and translation codes they contain."

All this matter is connected with the two origins of life of Freeman Dyson[4], the role of entropy in metabolism and with previously discussed concepts, namely:

    (a) the relational order of Stebbins[5];
    (b) the relationship between negentropy and photosynthesis;
    (c) the relationship between negentropy and information.

The deterministic view of chance and necessity (Monod, Nobel laureate in Medicine and Biology) becomes obsolete in the light of the thermodynamic view of self-organization (Prigogine, Nobel laureate in Chemistry).

In his book *"The organic codes. The birth of semantic biology"*, Barbieri[3] raises new suggestions as regards the role of different genetic codes in the history of biological evolution. Barbieri presents a mathematical model of epigenesis.

In the new paradigm of evolutionary physics, *epigenesis* is a *milestone* on the road concerning biological evolution and biophysical chemistry.

---

[3] M. Barbieri, *The Organic Codes. The Birth of Semantic Biology*, PeQuod, Ancona, 2001.

[4] F. Dyson, *Origins of Life*, Hardcover edition, Cambridge University Press, 1986 and Paperback edition, Cambridge University Press, 1999.

[5] G.L. Stebbins, *The Basis of Progressive Evolution*, University of North Carolina Press, Chapel Hill, 1969.

Barbieri underlines the importance of Eigen's (Nobel laureate in Chemistry) paradox:

> "in order to have protein enzymes it is necessary to have a large genome, but in order to have a large genome it is necessary that protein enzymes are already present"

and adds that this paradox

> "has turned out to be a formidable obstacle for all replication-first theories" and that "abandoning the replication paradigm does not mean underestimating the importance of replication: it only means that, to the best of our knowledge, biological replication could not appear at the beginning but only at the end of precellular evolution."

Barbieri's contribution to science is the introduction of the concept of semantic biology.

Before looking at semantic theory, it is worth underlining that the *dissipative structures* of Prigogine and the *autopoiesis* of Varela[6] are prerequisites for this view in biophysical chemistry. Prigogine and Varela never had the opportunity to meet and discuss their ideas, though they both contributed to the same international meeting organized by the University of Siena[7]. It is now possible to look at dissipative structures and autopoiesis as two faces of the same coin and to recall the definition Varela gave of life in 1996:

> "A physical system can be said to be living if it is able to transform external energy/matter into an internal process of self-maintenance and self-generation. This common sense, macroscopic definition, finds its equivalent at the cellular level in the notion of autopoiesis."

Returning to Barbieri the main point is the role of the *semantic processes*, related to codified *assemblies*.

---

[6] H. Varela and H. Maturana, *Autopoiesis and Cognition. The Realization of the Living*, D. Reidel Publishing Company, Dordrecht, Holland, 1980.

[7] Tempos in science and nature: structures, relations, and complexity, *Annals New York Acad. Sci.*, **879**, New York, 1999.

Barbieri claims:

"It is logical therefore to conclude that the informatic biology of the 20th century will have to be superseded by a more general biology that also takes into account the semantic processes of nature which are such an integral part of life.

A semantic view of life, a view that takes energy, information and meaning into account, must be able to offer alternative explanations in all fields, and in particular it must be able to propose new models of the cell, embryonic development and evolution."

According to Barbieri:

"The cell is an autopoietic and epigenetic system made of three fundamental categories (genotype, ribotype and phenotype) which contains at least one organic memory (the genome) and at least one organic code (the genetic code).

Semantic theory has its foundation in the idea that a cell is an epigenetic system, and states that organic codes and organic memories are indispensable precisely because only they can make a phenotype more complex than its genotype. If the living cell is an epigenetic system, then it is bound to have organic codes, and it is bound therefore to be a semantic system."

This is why pre-implant diagnosis of an embryo, obtained by taking two of the embryo's eight cells, completely upsets the information and meaning of that future life.

9.3. Epigenetic paradigm versus genetic paradigm

In her book *"Genetic Engineering. Dream or Nightmare?"*[8], Mae-Wan Ho discusses the epigenetic paradigm versus the genetic paradigm:

"History has a habit of creating heroes and anti-heroes: Darwin triumphed while Lamarck fell into ridicule and obscurity. This is because the theories of the two are diametrically opposed. The theory of Darwin is based on natural selection and selection means separation of an organism from its environment. In this way, the

---

[8] M.W. Ho, *Genetic Engineering Dream or Nightmare?* (2nd edn), Third World Network, Gateway Penang and Bath, Gill & Macmillan and Continuum, Dublin and New York, 1999.

organism is separated conceptually from its experience and the consequence is the barrier of Weismann and the central dogma of molecular biology, the objectives and practice of which are both reductionist. The vital experiences of the organism and the organism itself are both denied, since only its genes seem to have consequences for development and evolution. This fatalism is intrinsic in the deterministic genetic paradigm.

The theory of Lamarck is based on a concept of the organism as an active, autonomous being, open to its environment, during *epigenesis* or development. Since its openness is a threat to the *status quo*, it is not surprising that the theory was forgotten.

The theory invites us to look more closely at the dynamics of transformation and the mechanisms by which transformation can be "internalised" in the course of development and evolution. It is in line with the epigenetic approach that emerged as an alternative to neo-Darwinism at the end of the 1970s. This approach, to which I contributed, is embraced by the new genetics. The epigenetic approach regards experience of the environment during development as central in the evolution of the organism. This concept is still potentially subversive with respect to the *status quo*, which is why it is vehemently rejected by current orthodoxy."

Ho's point of view is perfectly in line with the epistemological basis of the First Step and with the role of epigenesis previously outlined. Moreover, the interplay between entropy and negentropy, between metabolism and replication, and Prigogine's thermodynamics are the scientific background of this vision. Again Mae-Wan Ho is far both from the ideology of creationism and of that of the neo-Darwinism.

9.4. Origins of life

Starting from the lessons of Schrödinger in Dublin[9] in 1943 and the theories of Eigen[10] and Oparin[11], and using the mathematics of von Neumann[12] and

---

[9] E. Schrödinger, *What is Life? The Physical Aspects of the Living Cells*, Cambridge University Press, Cambridge, 1947.

[10] M. Eigen, W. Gardiner, P. Schuster and R. Winckler-Oswatitch, The origin of genetic information, *Sci. Am.*, **244**, 88–92, 1981.

[11] A. I. Oparin, *The Origin of Life on the Earth*, Oliver and Boyd, Edinburgh, 1957.

[12] J. von Neumann, *Theory of Self-Reproducing Automata*, University of Illinois Press, Illinois, 1966.

the biology of Margulis[13], Dyson[4] constructed a model that envisaged the transition from chaos to a state of organized metabolic activity in a molecular population sufficiently rich to ensure homeostasis.

Dyson[4] makes the distinction between the mechanism of replication and that of metabolism. Both these phenomena, according to Dyson, have a conceptual basis in physics: replication by virtue of the quantum mechanical stability of molecular structures; metabolism by virtue of the capacity of a living cell to obtain negative entropy from its surroundings, in line with the laws of thermodynamics.

Today we know that life is a set of relations and coevolutions reaching us from distant biological eras; an infinite set of interactions between molecules and cells, between atmospheres and living beings, between biological species and ecosystems and we know that life is more characteristic of these systemic interactions than of a single individual. Lucretius was right in saying that life is passed from one being to the other, given in property to no one but in use to all.

Today we know that just as we cannot imagine metabolic life without replication, we cannot imagine replicative life without metabolism. Dyson was right in saying that defending biodiversity means opposing the destruction of the relations and histories that created differences. Replication alone means the risk of having many things that are all the same and with the same history. Metabolism, stochastic encounters with the environment and negentropy, however, play a historical-creative role in biological evolution. The new creativity is stabilized and transmitted by DNA.

Dyson identifies two origins of histories of life: that of proteins, the hardware of the computer that physically processes information, and that of nucleic acids, the sophisticated software that contains and transmits information. Life began twice, with different creatures, one capable of metabolism, the other of replication. They met, perhaps much later, a long time ago: a meeting favoured by chance, but less improbable than simultaneous origin.

In other words, if the spontaneous appearance of a protein structure is an improbable event in the midst of molecular chaos, if the spontaneous appearance of a nucleic acid structure is also unlikely, the two improbable events, Dyson emphasizes, are more likely to occur independently over a long span of time, than together. Hence the first organisms were probably cells with a metabolic apparatus controlled by proteins and devoid of genetic

---

[13] L. Margulis, *Origin of Eukaryotic Cells*, Yale University Press, New Haven, 1970.

apparatus: the first organisms consisting exclusively of proteins, could have lived independently for a long time, gradually developing a more and more efficient metabolic apparatus.

9.5. Metabolism is the origin of biodiversities

Histories, encounters and relations are favoured by time, real time, an intrinsic property of matter. Time moulds forms, leaving traces of encounters, the footprints of past relations. Nature, rich with information and evolutionary history, allows new encounters and new diversities.

According to Dyson, the creativity that unfolds almost by chance in complicated structures is more important as a moving force of evolution than Darwinian competition between replicating monads.

9.6. Entropy, information and the Maxwell's demon

The role of information is directly connected with the problem of Maxwell's demon. Brillouin[14] showed that the demon would require information about the molecules. More energy would be needed to obtain this information than the energy gained, which means that Maxwell's demon does not exist in a classical or physical system. The problem is completely different if we deal with living systems able to create information and self-organization and to decrease entropy.

Again the conclusion is that in a living system entropy (and negentropy) is not a state function (see Fourth Step).

According to Barbieri[3]:

"In the end, biologists will discover that organic codes and organic memories are the sole instruments that allow a system to increase its own complexity, and will understand that this is the most fundamental property of all living creatures, the very essence of life."

This also means that it is not possible to compel nature in a cage of mechanical laws, of aseptic mathematical models.

---

[14] L. Brillouin, *Science and Information Theory* (2nd edn*)*, Academic Press, New York, 1962.

"The rules of law are accessory, those of Nature essential; those of law are agreed upon, not native, those of Nature native, not agreed upon." (Antiphone the Sophist, 5th century BC)

9.7. The blue print

This paragraph emphasizes another important step written by Brooks and Wiley[15], related to the role of dissipative structures in nature.

"Our theory differs from other so-called thermodynamic theories in claiming that information takes precedence over energy when we consider the impact of the second law of thermodynamics on organisms. We can see this point when we consider the fact that it is the instructional information that determines the kinds of chemical reactions that will occur at a rate fast enough to maintain the steady state of an organism in a particular environment. In other words, instructional information is not only directly related to structural organization, it also determines how energy will flow through the organism. (...)

(...) since organisms are open thermodynamic systems, there are no particular energy constraints placed on increasing biological complexity and organization.

(...) partial entropy functions are associated with the genetic code and with other hierarchically organized aspects of the information systems of organisms.

(...) everyone agrees, we presume, that organisms are far-from-equilibrium dissipative structures.

(...) why there are millions of species of organisms but only a few kinds of steam engines?

(...) *organisms carry their blueprints inside and constantly refer to them, while the blueprint of a steam engine stays on the engineer's desk.*"[15]

---

[15] D. R. Brooks and E. O. Wiley, *Evolution as Entropy: Towards a Unified Theory of Biology*, The University of Chicago Press, Chicago, 1986.

9.8. History matters

Living organisms, and life itself, come into being through self-organizing processes, such as morphogenesis, epigenesis and metabolism, which generate order from disorder. One of the most striking feature of self-organizing systems is their ability to give rise to spatiotemporal patterns.

Archaea micro-organisms emerged about 3.5 billions years ago, and represent one of the oldest form of life. They live in extreme habitats, such as thermal vents or hypersaline water, which reflect the ancestral environment on Earth, but they are also abundant in the plankton of the open sea.

Archaea micro-organisms give rise to interesting patterns.

Patterns, processes, no mechanical molecules paved the way of the biophysical evolution.

"Then there is Francis Crick, who was so enamoured of the molecule he helped to discover that he simply could envision no earthly origin for it. It must have arrived from extra-terrestrial sources."[16]

History matters, as pointed out by Niles Eldredge[17]:

"But what I am saying in no uncertain terms is that simple extrapolation upwards of a competitive model for transmission of genetic information from one generation to the next does not suffice to explain more than a fraction of the commonly encountered patterns in the evolutionary history of life. The fault in this reductionist, extrapolationist vision lies in the penchant for "arm-waving," for seeing physical events as isolated incidents that set selection off in this direction or that. This tired old "in principle" explanation of how the evolutionary process works does not do the job. It does not explicitly link, in a general theoretical framework, evolutionary biological systems with the rest of the physical world. Indeed, to insist upon a form of internalized competition among genes themselves for representation in the next generation as the basic motor of evolutionary change is to substitute a pseudo-physical

---

[16] R.E. Ulanowicz, Process ecology: a transactional worldview, Moen Brainstorming Meeting and *Int. J. Ecodynamics*, in press, 2006.

[17] N. Eldredge, *The Pattern of Evolution*, W.H. Freeman and Company, New York, 1998.

model, while abandoning the search for law-like links to the actual physical universe.

(...) Now that plate tectonics has melded the study of the patterns of geological history with an understanding of the dynamic processes that produce those patterns; and now that progress has also been made in integrating patterns of biological history with theoretical understanding of the evolutionary process, the time has come to acknowledge and embrace the physical context of biological systems – and the evolutionary history of life."

Concluding this semantic discussion we may say that it is not possible to understand living systems only with mathematical sciences.

10
TENTH STEP

THE PROBABILITY PARADOX

What is man in nature?
Nothing compared to infinity,
everything compared to nothing,
something intermediate between all and nothing.
Infinitely far from understanding these extremes,
the end and the beginning of things are to him inscrutable.

Blaise Pascal[1]

10.1. The extinction of dinosaurs

In his lecture at Siena University on the occasion of his *laurea honoris causa*, Walter Alvarez, geologist and professor at the University of California at Berkeley, the scientist who proposed a new explanation for the extinction of dinosaurs, offered some interesting insights into the role of special events:

"In 1980, after a year of attempts to understand it, our Berkeley group proposed that the iridium had been brought in by a large comet or asteroid, whose impact on Earth had caused the giant KT (Cretaceous-Tertiary) mass extinction.

Other workers confirmed the iridium anomaly as a world-wide feature and began to find other evidence for the KT impact, including quartz grains with the damage produced by shock during

---

[1] B. Pascal, *Pensées,* Penguin USA reissue, 1995. In this edition, the *pensées* are nos. 164 and 223.

impact, and spherules representing droplets of impact melt that had
been blasted outside the Earth's atmosphere.
This changed some of the ways geologists think. Instead of
always looking *down* at the rocks, geologists began to look upward
at the sky, and became interested in comets and asteroids, which on
rare occasions produce big impacts on Earth."

Geologists could no longer doubt the reality of impact events.
Alvarez underlined the role of *impact events* and discussed the interplay of
cycles and the arrow of time.

"An early Italian geologist named Dante Alighieri knew about one
kind of geological cycle – the cycle in which water is evaporated
from the sea, falls as rain, and flows back down rivers to the sea. In
Canto 14 of *Purgatorio*, he applied this concept to the Arno…

"… ché dal principio suo, ov'è sì pregno
l'alpestre monte ond'è tronco Peloro,
che 'n pochi luoghi passa oltre quel segno,
infin là 've si rende per ristoro
di quel che 'l ciel della marina asciuga,
ond'hanno i fiumi ciò che va con loro…"

Cycles are fundamental to geology.
In Earth history we also see trends – what we might call *arrows*,
in contrast to cycles. The cycling of compression and extension in
Italy is generating the Apennines – an arrow of Earth evolution that
cannot be undone. And the rock cycle, in the broader context of the
supercontinent cycle is extracting magma from the deep earth and
slowly converting it into sandstone and clay, another great arrow that
cannot be reversed.
Philosophers of *human* history have long debated whether history
is fundamentally cyclical or fundamentally arrow-like. Based on
what we see in the sister field of geology, I suspect that neither
cycles nor trends are alone responsible for the patterns of history, but
rather they work hand in hand, and that great cycles, like the rise and
fall of civilizations, mediate long-term trends, like the growing
complexity and interconnectedness of human societies all over the
Earth.

Maybe there is a further lesson from geology. Every now and then the continuing interplay between cycles and arrows is broken by a great *event* that fundamentally changes the direction of Earth history in an absolutely *unpredictable way*. This was the role of the Cretaceous-Tertiary extinction. Prior to that event – the chance impact of a comet or asteroid – dinosaurs had ruled the earth for almost 200 million years, and mammals were condemned to a marginal role as small creatures. We are the dominant species on Earth today only because that great event removed the dinosaurs.

Is there an analogy in human history? I think so. Historians have often considered the importance of the "Great Man" – someone whose thoughts or deeds deflected human history into unexpected, unpredictable directions.

Cycles, arrows, and great men or great events? Here is a profound concept for historians of the Earth and of humanity to explore together. But it is only the beginning. Historians and philosophers have debated for more than two thousand years whether the future is predetermined or whether it can be directed into particular paths by human free will. The mathematics of chaos theory and sensitive dependence on initial conditions has proven, I think, in the last two decades, that the *future cannot be predetermined*. It is contingent – it is sensitively dependent on a myriad of *unpredictable chance events*, like the Cretaceous-Tertiary impact. What a concept to explore in both of our branches of history!

Universities like Siena and Berkeley are places for thinking, talking and writing about those concepts – concepts like "time", "life", "history", "God", "justice", "art" and "music".

As humans, we no longer have to operate by the merciless criterion of natural selection. Instead, we can do many experiments, harmlessly, in our minds and choose the outcome that we think seems most desirable."

10.2. Probability and prediction of extreme events

An extreme event is generated by a system with complex dynamics entering an extreme state. This definition implies that extreme states are rare, since otherwise they would contribute so prominently to the variance that their distance to the mean would be of the same order of magnitude as the variance. Hence, an extreme state is a state that is far from the region in state-space where the system is *normally*, and its occurrence in time is irregular and seemingly random[2].

We call a system complex, when its dynamics is complex, irrespective of whether a mathematical description (or an experimental realization) of a system is simple or complicated.

Complex dynamics means that the time evolution is irregular and difficult to predict, but nonetheless contains structures in time or space that emerge without being designed[3]. Examples include classical chaos[4], spatiotemporal chaos and hydrodynamic turbulence[5], when we consider perfectly deterministic systems, and physical growth processes[6] and systems exhibiting self-organized criticality[7] as stochastically driven systems.

A common characteristic of these systems is that their distribution in phase space (for those where it is defined) or probability distributions for the magnitude of typical observables has thin parts or tails. Dynamically speaking, instability of solutions, positive nonlinear feedback loops and the lack of conservative quantities enable such systems to generate extreme events, i.e. to make large excursions from the mean state of the system[8].

Predictability of a given phenomenon means that there are temporal correlations in time series data that in principle allow predictions to be made with a certain accuracy. In other words, the average prediction error is limited from below by the dynamics that can be quantified by the information-theoretic concept of Kolmogorov–Sinai entropy.

An extreme event is usually of very large magnitude. Hence, a prediction error of a given size, which might be so big that it does not help for the prediction of *normal* events, might still be small compared to the magnitude of an extreme event never observed before and hence might not be helpful for its prediction.

What is *extreme* clearly depends on the point of view, and, in particular, on the observable. Generally speaking, an extreme event is one that is far

---

[2] H. Kantz, Extreme events in nature – a challenge for the understanding of complex dynamics, *Int. J. Ecodynamics*, 2006.

[3] R. Badii and A. Politi, *Complexity*, Cambridge University Press, Cambridge, UK, 1999.

[4] H.G. Schuster, *Deterministic Chaos*, VCH-Wiley, 1995.

[5] U. Frisch, *Turbulence*, Cambridge University Press, Cambridge, UK, 1995.

[6] P. Meakin, *Fractals, Scaling and Growth far from Equilibrium*, Cambridge University Press, Cambridge, UK, 1998.

[7] P. Bak, C. Tang and K. Wiesenfeld, Self organized criticality: an explanation of 1/f noise, *Phys. Rev. Lett.*, **59**, 381, 1987.

[8] H. Kantz and M. Ragwitz, Phase space reconstruction and nonlinear predictions for stationary and nonstationary Markovian processes, *Int. J. Bifurcation Chaos*, **14**, 1935–1945, 2004.

from the mean of the distribution. More specifically, in order to exclude the vagueness of common language, we must define the notion of an extreme state[9].

An extreme state is a state that is far from the region in state space where the system is normally, and its occurrence in time is irregular and seemingly random[8].

From the point of view of nonlinear dynamics, Kantz[9] proposed prediction based on hidden Markovian models as a nonlinear stochastic process and on an alternative formulation of the Langevin equation[10]:

$$\dot{V} = -\gamma V + L(t).$$

The right-hand side is a force consisting of a linear dumping term in V with a constant linear coefficient γ, plus a term $L(t)$, which can be treated as a stochastic process with $\langle L(t) \rangle = 0$ and $\langle L(t) \, L(t') \rangle = \Gamma \delta(t - t')$, were Γ is a constant.

The main problem remains almost the same: *previously unobserved events cannot be predicted.*

10.3. The chameleon effect

An event occurs in a stochastic manner because it is preceded by others. There are genetic and environmental constraints. Evolutionary events proceed in a manner that depends on time: they show a direction of time; they are irreversible. Past time has determined the constraints; the future is largely unpredictable, and always has a stochastic or probable element.

Previously unobserved events cannot be predicted; rare and extreme events may completely change the dynamics of complex systems.

Fig. 27 shows the *emergence* of a probability paradox in the presence of events.

---

[9] H. Kantz, D. Holstein, M. Ragwitz and N.K. Vitanov, Markov chain model for turbulent wind speed data, *Physica A*, **342**, 315–321, 2004.

[10] N.G. van Kampen, *Stochastic Processes in Physics and Chemistry*, North Holland, Amsterdam, 1992.

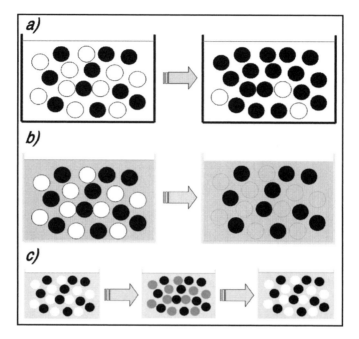

Figure 27: Unexpected events that may occur in living systems: (a)
oxidation, (b) chameleon effect, (c) oscillating reaction.

(a)   Unknown to the observer, oxidation (a chemical event) occurs
      in the classic "white and black spheres" game: the white–black
      probability is no longer fifty–fifty (only if oxidation changes
      white spheres to grey can we know what happened).

(b)   An evolutionary event related to the *"chameleon" effect*
      (sensitivity to the environment) occurs: again the probability is
      no longer fifty–fifty and the interval of the event depends on the
      "chameleon".

(c)   An oscillating event similar to the Belousov–Zhabotinsky
      reaction (see Fifth Step) occurs: the situation is more complex
      and depends on many parameters. Again the observer cannot
      predict which sphere will be picked out of the container.

It is possible to conclude that in the far-from-equilibrium framework, a
classical probability approach does not apply and new models should be
developed for the Boltzmann relation $S = k \ln W$.

Recently Nielsen and Ulanowicz[11] discussed the concept of probability in the light of Popper's last book.

"The late K.R. Popper in one of his last books[12] introduced a new world view. The world is seen, not in terms of deterministic forces, but rather as a world of propensities. Observed phenomena are considered to be outcomes of coinciding events that possess non-equiprobable distribution. That is, the world behaves not like the toss of a coin or a pair of dice, where all outcomes have the same probability of 1/2 or 1/6, respectively. Rather, it behaves more like a game where the coins or dice are loaded. Furthermore, the probabilities themselves can change as phenomena interact with each other, i.e. the probabilities themselves are not stationary. Any probability becomes conditional upon surrounding events. This represents a remarkable change from previous Popperian philosophy, which was dominated by a deterministic and reductionistic world view. It would seem that late in his life Popper has renounced a fixed world in favour of one that is clearly non-deterministic and even holistic in character. The propensity world view recently has influenced a reinterpretation of the ascendency principle originally formulated by Ulanowicz[13,14,15]."

This is perfectly in agreement with our epistemological view.

10.4. Ageno: the origins of irreversibility

Mario Ageno, the late professor of Physics at the University of Rome (La Sapienza), showed an interesting new point of view on probability and uncertainty in his book *The origins of irreversibility*[16].

---

[11] S.N. Nielsen and R.E. Ulanowicz, On the consistency between thermodynamical and network approaches to ecosystems, *Ecological Modelling*, **132**, 23–31, 2000.

[12] K.R. Popper, *A World of Propensities*, Thoemmes, Bristol, 1990.

[13] R.E. Ulanowicz, *Growth and Development. Ecosystems Phenomenology*, Springer, Berlin, 1986.

[14] R.E. Ulanowicz, The propensities of evolving systems, in: E.L. Khalil and K.E. Boulding, eds., *Evolution, Order and Complexity*, Routledge, London, 1996.

[15] R.E. Ulanowicz, *Ecology, the Ascendent Perspective*, Columbia University Press, New York, 1997.

[16] M. Ageno, *Le origini della irreversibilità*, Bollati Boringhieri, Torino, 1992.

"At each collision of a molecule against the walls, there is a non zero probability of transition to any of a continuum of microstates, and due to the unstable character of the trajectories in phase space, all microstates possible *a priori* for the gas therefore become accessible. This is the only reason why most times the gas is observed and for most of the time it is observed, it is found to be in that phenomenological (macroscopic) state that covers a much greater number of microstates than all other states of the kind, namely, the state of thermodynamic equilibrium. If our gas is going through an obligatory succession of microstates, $b_i \rightarrow b_{i+1} \rightarrow b_{i+2} \rightarrow ...$, the transition from one microstate to the next is caused by an elastic collision between two molecules. When the gas is in microstate b_{i+k}, there comes a moment when one of its molecules collides with a wall of the container. The inelastic collision causes a transition to another obligatory succession of microstates, $c_1 \rightarrow c_2 \rightarrow ...$, completely independent of the former succession and microstate c_1 will be chosen randomly on the basis of the laws of probability from a continuous range of possible microstates. The most probable choice is a microstate belonging to the macrostate richest in microstates in the range.

If we now imagine inverting the direction of all molecular velocities while the gas goes through the obligatory succession of microstates b, the succession will run in the opposite direction $b_i \rightarrow b_{i-1} \rightarrow b_{i-2} \rightarrow ...$. However, there comes a time when the transition from a certain microstate b_{i-b} to the next is caused by collision of the molecule against the wall. Again, this causes a transition to another obligatory succession of microstates, $a_n \rightarrow a_{n-1} \rightarrow a_{n-2} \rightarrow ...$, quite independent of succession b, and now microstate a_n is chosen at random from the continuous range of alternative possibilities. Once again, the most probable choice falls on a microstate belonging to the macrostate richest in microstates in that range. We see that in whatever direction the gas goes through part of the obligatory succession of microstates, collision of one of its molecules against the walls will always tend to make it go from the macrostate in which it was before the collision to another macrostate richer in microstates. This transition is the most probable event when the gas leaves the obligatory succession in which it was. By frequent collisions of molecules against the walls, most of the times we observe the gas and for most of the time we observe it, the gas will therefore be found in the macrostate richest in microstates, namely the state of thermodynamic equilibrium.

According to the paradigm of gases, we see that the laws governing elementary processes can identify a linear order of events occurring at that level but cannot attribute a direction to the linear order. In the case of the deterministic model gas of Boltzmann, the laws of mechanics enable us to establish that the three microstates *a, b* and *c* belong to the same obligatory succession of the gas. For example, microstate *b*, is between states *a* and *c*, but the laws cannot tell us which is first and which later, whether *a* precedes *c* in time or whether *c* precedes *a* in time. According to the laws of mechanics, there is perfect symmetry between the two cases.

Now, the event that separates the time spent in a microstate from that spent in the adjacent microstate is in any case a molecular collision. There are always two collisions at the two limits of the time the gas spends in a given microstate. If one is a collision with the walls of the container, there is a transition from one obligatory succession of microstates to another quite independent succession immediately adjacent to it in time on both sides.

In the first collisions with the walls in a given order, we may find that the previous macrostate was particularly poor in microstates and becomes richer as a result of the collision. Collision after collision, the gas on average achieves richer and richer macrostates, each time choosing between the most probable available. This means that the gas evolves towards a state of equilibrium: the order arbitrarily chosen by us in the chain of microstates is the one that will most probably occur physically and it is indicated by the arrow of time.

If, on the other hand, in the order we arbitrarily chose, we find that the gas on average goes to increasingly improbable macrostates, increasingly poor in microstates, from collision to collision with the walls, clearly that order has no appreciable probability of occurring physically in the segment of the chain of microstates we considered: the arrow of time points in the other direction.

It is therefore the continuous character of the range of choices open to the gas at every collision that gives rise to asymmetry between the two directions in which the chain of microstates is followed: a probability, proportional to the number of microstates in each available macrostate, is attributed to the two directions each time. However, the continuity of the range of choices springs in turn from the probabilistic character of the phenomena of absorption and emission of electromagnetic radiation by the gas. The irreversibility of the macroscopic processes therefore originates in the uncertainty between energy and time.

The new interpretative framework of the second law of thermodynamics and the origins of irreversibility of macroscopic processes, with recognition of the stochastic nature of the inevitable residual "perturbations", finally offers a reasoned foundation for the use of probabilistic methods in statistical mechanics and in classical physics in general."

This aseptic concept of probability does not consider the capacity of living systems to self-organize. However, the role of collisions against the walls is important.

Ageno[16] is only wrong when he writes: "Clearly that order has no appreciable probability of occurring physically in the segment of the chain of microstates we considered: the arrow of time points in the other direction".

As we have seen in the previous chapters, it is the temporal sequence of the events, which determines the arrow of time: this cannot be reversed.

Moreover, in the liquid state, self-organization (see Fifth Step) leads to ordered macrostates poor in microstates.

11
ELEVENTH STEP

ALEA IACTA EST

An idea that is not dangerous isn't an idea at all.
Coherence is the last refuge of those without imagination.

Oscar Wilde

11.1. Crossing the Rubicon

When Julius Caesar decided to cross the Rubicon, declaring "*Alea iacta est*", he knew he was breaking the rules of Roman power. In science, crossing the Rubicon is a metaphor for a change of paradigm and the new always has many enemies: power holders, conservative moralists (Cicero) and demagogues (Brutus). The funeral oration of Mark Antony in Shakespeare's Julius Caesar explains this. If the Rubicon had not been crossed and if Mark Antony's speech had not been made, the civilizations of Augustus, Tiberius, Marcus Aurelius, Hadrian, Trajan and Constantine would not have existed.

The Rubicon is a metaphor for the entropy watershed between the living and non-living. To cross it, scientific (Prigogine), epistemological (see Prologue and First Step) and semantic (Ninth Step) foundations are necessary. It is necessary to have reference values (e.g. biodiversity) and the right scientific tools (e.g. goal functions) and it is also necessary to have the humility and courage to build a new science. At the moment, we can only see the outlines of this new science.

The first thing to do is to build a pontoon. The wooden boats that compose it are the models and experiments that enable us to glimpse the behaviour of living things, but are not yet "evolutionary physics". We saw at least two of these boats in the Fifth Step: the supramolecular structure of water and

Belousov–Zhabotinsky reactions. Although they are incomplete models, they are useful for the crossing towards evolutionary physics.

Now comes the challenge for mathematicians, physicists, physical chemists and ecologists. The challenge will be played on the fields of temporal logic, probability, uncertainty, Gödel's theorem and Poincaré's theories. It will exploit narrative and qualitative elements: in evolutionary physics, reproducible experiments do not exist. The challenge will take the courage to sail in the open sea, far from the calm harbour of determinism, mechanism and reductionism, towards a "gaya" and aleatory science. The challenge is to build a new physics and a new chemistry of the living world, evolving systems, the whole of nature. The two fundamental categories for this scientific challenge are time and aesthetics.

11.2. Time's revolution

The main theme of the scientific work of Ilya Prigogine was the role of time in the physical sciences and biology. He contributed significantly to the understanding of irreversible processes, particularly in systems far from equilibrium. The results of his work on dissipative structures have stimulated many scientists throughout the world and have profound consequences for our understanding of biological and ecological systems.

In his Nobel Lecture (8th December 1977) Prigogine stated:

> "The development of the theory permits us to distinguish various levels of time: time as associated with classical or quantum dynamics, time associated with irreversibility through a Lyapounow function and time associated with 'history' through bifurcations. I believe that this diversification of the concept of time enables a better integration of theoretical physics and chemistry with disciplines dealing with other aspects of nature."

As only a very intelligent person can do, he wrote an *autobiography*, which, like his scientific career, was distinguished by a total absence of hypocritical respect. Here are some particularly significant passages:

> "… Since my adolescence, I have read many philosophical texts, and I still remember the spell *"L'évolution créatrice"* cast on me. More specifically, I felt that some essential message was embedded, still to be made explicit, in Bergson's remark: *"The more deeply we study the nature of time, the better we understand that duration*

means invention, creation of forms, continuous elaboration of the absolutely new."

Fortunate coincidences made the choice for my studies at the university. Indeed, they led me to an almost opposite direction, towards chemistry and physics. And so, in 1941, I was conferred my first doctoral degree. Very soon, two of my teachers were to exert an enduring influence on the orientation of my future work."

This is the first challenge: the irreversible character of time.

Ecodynamics or ecological thermodynamics has to be based on a rigorous mathematical approach, but the equations have to be irreversible with respect to time, not based on a rigid intellectual cage (framework) but open enough to contain narrative elements. A temporal-logic approach is necessary because the slight movement of a butterfly's wing may cause a cyclone in the Azores. To save the planet, to avoid another New Orleans, models must capture the future: the key is the sequence of rare events.

11.3. The role of aesthetics

"The limits of a purely quantitative view of nature, which denies the fundamental ecological category of quality and the importance of the aesthetic element, are becoming evident in the face of the complex temporal dynamics of the biosphere and global ecosystem. These temporal dynamics are based on multiple relations in a process of coevolution based on forms, colours, sounds, odours and flavours."[1]

The history of nature is a systemic and evolutionary history, one in which quantity and quality are ever present, a story in which the aesthetic element plays a determinant role.

Observation of nature tells us two important things: that quality and time are not external values but intrinsic properties of living organisms. This is the great lesson of Darwin's theory of evolution, a theory which, among other things, has the merit of not indicating aims or certainties for evolution. Darwin repeatedly underlined the fundamental role of chance and the absence of an "end" towards which life as a whole moves.

In this context, time modulates forms and structures, sounds and colours: these properties run through the entire history of evolution. How then can

---

[1] E. Tiezzi, *Beauty & Science*, WIT Press, Southampton UK, 2005.

western science remain anchored to a merely geometric reading of nature, a mechanistic conception of physical laws?

An example of these "absolute laws of nature", that Prigogine contrasts with Whitehead's concept of *event* ("an event cannot be reduced to a point in the four-dimensional continuum", the space-time of relativity), are Newton's laws of mechanics and gravitation. These laws are deterministic and reversible; they assume symmetry of time between past and future. This is not true of living systems, ecosystems, biological and ecological events.

Obviously the mathematical machine *par excellence*, the computer, cannot understand the concept of evolution, the arrow of time. As with all machines, it is indifferent to the irreversibility of time, incapable of understanding the real meaning of time.

Recently some studies in mathematical logic have examined the possibility of getting computers to understand the concept of the passage of time. Indeed, the study of real-time systems, in other words systems in which temporal evolution plays a primary role, has made interesting advances. Specifically, "the properties to describe in these systems are not only qualitative, properties which classical temporal logic can express, but also quantitative"[2].

It would be interesting to develop logics that express "eternal" constraints, such as the three dimensions, on one hand, and that tackle the real meaning of evolution, and hence the importance of *events* and their *successions*, on the other.

Nature is evolutionary in character. The more one seeks to comprehend her, in the etymological sense of *enclosing, imprisoning*, in our mental schemes, the more she creates relations and complexity, memories and creative possibilities. It is the passing of time that prevents us from capturing the fleeting moment of global knowledge.

Dostoyevski's idiot sustained that beauty would save the world. In Phaedrus, Plato writes that "Beauty shone up there like Being, and we down here perceived it with our clearest senses [...]: only Beauty received this most manifest and lovely being".

Nature is a work of art. To understand this, nature must be viewed whole like the fresco of Bergson, contemplated with the eyes of our culture and our senses: heard, felt, thought – letting her fabric and rhythms penetrate us.

Follow the meanders of connecting structures: now the mauve and turquoise of the kingfisher, now the forms of a stone and a beautiful woman,

---

[2] El.B. Tiezzi, *Problemi di assiomatizzabilità per logiche real-time*, Doctorate thesis, Siena University, 1998.

now the song of the brook and the rustling of wind, now the perfume of sandalwood and the penetrating odour of resin, now the taste of balsamic vinegar and the aftertaste of Gewurz–Traminer – letting ourselves be inebriated by the five senses, thinking.

Beauty has left many traces in the great landlocked sea, on the coasts and in the interior of Mediterranean countries. It is transgressive beauty, just as scientific research should be. If everything is levelled, standardized, catalogued, weighed and measured with impact factors, economic indicators and iron laws, neither beauty nor science will remain.

On the other hand, the meeting of beauty and science, a strong manifest beauty and a science with the courage to abandon crude schemes and old paradigms, can give birth to "biodiverse" thought that can open doors and write stories that are always new.

11.4. The laws of ecosystem theory

Sven Jørgensen[3] reformulated the ecosystem theory in a compressed form with the following eight laws.

(1a) Mass (for instance, accounted as the elements) is conserved in ecosystems.

(1b) Energy is conserved in ecosystems.

(2) All processes in ecosystems are irreversible (this is probably the most useful way to express the Second Law of Thermodynamics in ecology). Another useful formulation, where the second law is applied to an ecosystem, is: ecosystems are driven by an input of low entropy energy, which after being used to maintain the ecosystem, is transmitted to the environment as high entropy energy.

(3) At 0 K, neither disorder nor order (structure) can be created. At increasing temperature, the processes creating order (structure) occur more rapidly, but the cost of maintaining the structure in the form of disordering processes gets higher. Carbon-based life is therefore found at temperatures where there is a good balance between the rates of these two opposite processes, i.e. at about 250–350 K.

---

[3] S.E. Jørgensen and Y.M. Svirezhev, *Towards a Thermodynamic Theory for Ecological Systems*, Elsevier, Amsterdam, 2004.

(4)   When a system receives a through-flow of exergy, after recovering energy for its maintenance, the system uses the exergy to move away from thermodynamic equilibrium. If a number of possibilities are offered, the one that moves the system farthest away from thermodynamic equilibrium is selected.

(5)   The growth of an ecosystem is possible by an increase in physical structure (biomass), by an increase in network (more cycling) and by an increase in information embodied in the system. All three forms of growth imply that the system is moving away from thermodynamic equilibrium and are associated with an increase in (1) exergy stored in the ecosystem, (2) through-flow (power) and (3) ascendency. An ecosystem receiving solar radiation aims for maximum exergy storage, maximum power and maximum ascendency.

(6)   The carbon-based life on the Earth has a characteristic basic biochemistry, shared by all organisms. This implies that many biochemical compounds can be found in all living organisms. They therefore have almost the same elementary composition and the composition of all organisms can be represented by a relatively narrow range of about 25 elements.

(7)   Biological systems are organized in a hierarchy. The variables describing the state at any level are determined by the processes at the level immediately below. In this way, each level has emerging properties.

Bernie Patten[4] added some interesting considerations.

"Systems in the abstract are most commonly defined as partially interconnected sets of components. Physics adds openness and that gives the under-treated concept of environment. Network aggradation, defined as a gain of internal order (negentropy) in excess of generated disorder (entropy) in movement away from thermodynamic equilibrium, accepts the dissipative principle that self-organization represents the balance between two opposing tendencies – movement away from thermodynamic equilibrium,

---

[4] B.C. Patten, Holoecology. Thirteen cardinal hypotheses reflecting the deep and essential holism of ecosystems and the living things they enfold within them, Moen Brainstorming Meeting, 2005.

aggradation, a property of input environs, and slippage back towards it, dissipation, occurring in output environs. Cycling involves both environs and so has both dissipative and aggradative sides. The balance, at ecological and human scales, is positive by the cardinal hypothesis, aggradation > dissipation, and hence the growth of order in the biosphere exceeds the production of disorder. The source of all this is a single ubiquitous property, so elementary as to escape attention – electromagnetic interaction. Conservative energy–matter transaction is the single force that drives organization and the origins of order follow. Existing far-from-equilibrium theory appears to have the correct phenomenology, but the missing element is how aggradation, potentially without bound, develops automatically as interactions increase, also potentially without bound. Network organization, and in particular network aggradation, would appear therefore to be at the heart of any future far-from-equilibrium ecological thermodynamics."

Robert Ulanowicz[5] recently made a very interesting epistemological synthesis.

" "Process ecology", depicts ecosystem development as arising out of at least two antagonistic trends via what is analogous to a dialectic: one direction is the entropic tendency towards disorganization and decay, which can involve singular events that defy quantification via probability theory. Opposing this ineluctable drift are self-entailing *configurations of processes* that engender positive feedback or autocatalysis, which in turn impart structure and regularity to ecosystems. The status of the transactions between the two trends can be gauged using information theory and is expressed in two complementary terms called the system "overhead" and "ascendency", respectively. Process ecology provides an opportunity to approach some contemporary enigmas, such as the origin of life, in a more accommodating light.

A subject discussed all too infrequently in ecological discourse is how the worldview that orients ecosystems science tacitly owes much to issues in classical philosophy. One of the very earliest splits in Western philosophy, for example, was between the *Eleatic* and

---

[5] R.E. Ulanowicz, Process ecology: a transactional worldview, Moen Brainstorming Meeting and *Int. J. Ecodynamics*, in press, 2006.

Milesian schools of Hellenistic thought. The central personality in the *Eleatic* School was Plato, who taught that the world is composed of eternal and unchanging "essences". What one perceives as changes are considered illusory distractions that impede the recognition of underlying essences. The major figure in the *Milesian* School was Heraclitus, whose famous quotation "Παντα ρει" infers that the only constant in the world is change. That one can never step twice into exactly the same river is a common exemplar of this dictum.

The second law of thermodynamics is decidedly a *Milesian* statement, and it has contributed to a number of hypotheses for ecosystem development.

Process ecology, the notion that objects are created by configurations of processes, provides a far more consistent framework for supporting the origin of life. At the same time process ecology provides a new avenue for further research on the origin of life."

Ecosystem science, like the world it describes, is in the process of becoming (Prigogine[6]).

11.5. Conclusive remarks

We may conclude that in classical science:

- geometric rules and mechanistic laws apply;
- Newton's laws are reversible deterministic laws.

Prigogine adds and counterpoises the concept of "events" to "laws of nature" of this kind. We know that the classical laws are not true for living systems, ecosystems and the events of biology and ecology.

As previously underlined (Prigogine), far from equilibrium we witness new states of matter having properties sharply at variance with those of equilibrium states. This suggests that irreversibility plays a fundamental role in nature.

We must therefore introduce the foundations of irreversibility into our basic description of nature (evolutionary thermodynamics).

---

[6] I. Prigogine, *From Being to Becoming*, W.H. Freeman and Co., San Francisco, 1980.

It is also important to underline that:

- *Space, by its structure, is reversible;*
- *Time, by its structure, is irreversible.*

In order to achieve an ecodynamic description we need to shift our attention from state functions to goal functions and to configurations of processes.

We may also underline the following two statements by Jørgensen[3]:

"The presence of irreducible systems is consistent with Gödel's theorem, according to which it will never be possible to give a detailed, comprehensive, complete and comprehensible description of the world. Most natural systems are irreducible, which places profound restrictions on the inherent reductionism of science.

Many ordered systems have emergent properties defined as properties that a system possesses in addition to the sum of properties of the components: the system is more than the sum of its components. S. Wolfram[7] (1984) calls these *irreducible systems* because their properties cannot be revealed by a reduction to some observations of the behaviour of the components."

In 1931, the young Viennese Kurt Gödel published a brief memoir on "formally undecidable propositions of *Principia mathematica* and similar systems", which concerned the incompleteness of a large class of formal theories, including arithmetic, as well as the impossibility of proving their coherence from within the theories themselves. Gödel's[8,9] theorem is often summarized as: "there is at least one formula of arithmetic that cannot be demonstrated" with the following formula:

$$(\exists\, y)(x) \sim Dim(x, y).$$

Interpreted in meta-mathematical language, the formula says "there is at least one formula of arithmetic for which no sequence of formulae constitutes a demonstration".

---

[7] S. Wolfram, Cellular automata as models of complexity, *Nature*, **311**, 419–424, 1984 and Computer software in science and mathematics, *Sci. Am.*, **251**, 140–151, 1984.

[8] E. Nagel and J.R. Newman, *Gödel's Proof*, New York University Press, New York, 1958.

[9] J.-Y. Girard, *Le champ du signe ou la faillite du réductionnisme*, in: E. Nagel, J.R. Newman, K. Gödel and J.-Y. Girard, *Le théorème de Gödel*, Editions du Seuil, Parigi, 1989.

Jørgensen[10] and Wolfram[7] underline that Gödel's theorem requires that mathematical and logical systems (i.e. purely epistemic, as opposed to ontic) cannot be shown to be self-consistent within their own frameworks but only from outside. A logical system cannot itself (from inside) decide on whether it is false or true. This requires an observer from outside the system, and this means that even epistemic systems must be open.

The impossibility of completely knowing the world is linked to the principle of Pascal, according to which the whole is more than the sum of its parts. This deals a heavy blow to reductionism. In chemistry and physics, there is no approach similar to that of Gödel, and there probably cannot be: however, there are some interesting analogies.

The first is Heisenberg's uncertainty principle and its thermodynamic extension (see Third Step). A first level of uncertainty concerns the relation between the measurer and the measurement and the impossibility of simultaneously determining the position and velocity or time and energy of a particle. There is also an intrinsic thermodynamic uncertainty of systems, which concerns the simultaneous presence of conservative and evolutionary quantities in living systems (see Third Step); the flow of time makes it impossible to grasp and measure the instant or other quantities that vary with the passage of time.

The second regards the probability (see Tenth Step) and impossibility of predicting anything if novelties emerge and unpredictable events occur within the system.

The mutual irreducibility of space and time makes it impossible to completely know living evolving systems.

Our wishes to the new science of evolutionary physics require a special toast, a narrative toast. To enjoy a good wine, one first observes its colour in the wineglass, then one savours its bouquet. Then two more senses are added: the tactile sense of the tongue to test its body and the palate to appreciate its flavour. The sense of hearing is missing and this is why glasses are touched.

True scientific knowledge should make simultaneous use of our minds and our five senses, as in the cognitive process of biological evolution.

We are children of the coevolutionary history of nature and man. To feel nature means to draw from the book of the billion-year experience of Mother

---

[10] S.E. Jørgensen, Toward a thermodynamics of biological systems, Moen Brainstorming Meeting and *Int. J. Ecodynamics*, **1**, 9–27, 2006.

Earth. To have knowledge – scientific and artistic – we need "ecstatic passion for the fabric of nature"[11].

If we really want to walk the path between aesthetics and science, throw out the paradigms of rigidity, welcome time into science and art, we must cast free from the arrogant scepticism of many scientists and have the courage to explore new forms of knowledge.

It is an attempt to wed the creative potential of art and science, aiming for a symbiosis of instinct and reason, a holistic approach to knowing nature that draws on the billion-year experience of other traditions.

---

[11] F. Novalis (F.L. von Hardenberg), *The Novices of Sais*. Original: *Die Lehrlinge zu Sais*, 1789. The work is an unfinished symbolic novel of the German Romanticism: Friedrich von Hardenberg, alias Novalis, died of tuberculosis at 29 years of age. Sais was a major religious and cultural centre near Alexandria in ancient Egypt.

NAMES INDEX

Beauty and Science

E. TIEZZI, University of Siena, Italy

In this book, the distinguished author argues that the aim of science should not be to dominate nature but to live in harmony with it. If we do not make reference to our "common biological origin" (Jean-Paul Sartre) or find the umbilical cord that binds us to nature, his conviction is that we risk destroying the life cycles of our planet. He demonstrates that the role of form, colour, flavour, sound, scent – and beauty – was fundamental for biological evolution, and is still fundamental today for a scientific view of complexity. This is especially so with nature threatened by the linear, mechanistic, arrogant and crude approach of science at the service of a society that "knows the price of everything and the value of nothing".

In order to avert catastrophe, Tiezzi asserts, science cannot be based only on reason but must combine reason, passion, intuition, emotion, logic and "global feeling": science cannot be cold.

Series: The Sustainable World, Vol 10
ISBN: 1-85312-740-X 2004 132pp
£39.00/US$62.00/€58.50

The End of Time

E. TIEZZI, University of Siena, Italy

"...a compelling fusion of the historical, philosophical and scientific aspects of the struggle towards a new, ecological culture."
ECOFARM & GARDEN

A best seller in Italy, this influential title has now been revised and translated into English for the first time. Tiezzi emphasises the need to reconcile the wants and pace of a modern generation with the hard reality that evolutionary history had already pre-determined a pace of her own. Presenting scenarios of 'hard' and 'soft' sustainability for the future, he poses the critical question: Will the scientific and cultural instruments we have be enough to combat the pressures of unsustainable human behaviour?

Series: The Sustainable World, Vol 1
ISBN: 1-85312-931-3 2002 216pp
£49.00/US$75.00/€73.50

We are now able to supply you with details of new WIT Press titles via E-Mail. To subscribe to this free service, or for information on any of our titles, please contact the Marketing Department, WIT Press, Ashurst Lodge, Ashurst, Southampton, SO40 7AA, UK
Tel: +44 (0) 238 029 3223
Fax: +44 (0) 238 029 2853
E-mail: marketing@witpress.com

WIT Press is a major publisher of engineering research. The company prides itself on producing books by leading researchers and scientists at the cutting edge of their specialities, thus enabling readers to remain at the forefront of scientific developments. Our list presently includes monographs, edited volumes, books on disk, and software in areas such as: Acoustics, Advanced Computing, Architecture and Structures, Biomedicine, Boundary Elements, Earthquake Engineering, Environmental Engineering, Fluid Mechanics, Fracture Mechanics, Heat Transfer, Marine and Offshore Engineering and Transport Engineering.